INTERNATIONAL
HARVESTER

INTERNATIONAL 1420

RED COMBINES

1915–2015

THE AUTHORITATIVE GUIDE TO INTERNATIONAL HARVESTER AND CASE IH COMBINES AND HARVESTING EQUIPMENT

LEE KLANCHER

GERRY SALZMAN, KEN UPDIKE,

MATTHIAS BUSCHMANN, JEAN COINTE, JOHANN DITTMER,

GREGG MONTGOMERY, MARTIN RICKATSON, SARAH TOMAC

Octane Press, First Edition
September 2015

Standard Edition
ISBN: 978-1-937747-45-9

Collector's Edition and Serial Numbered Collector's Edition
ISBN: 978-1-937747-58-9

Library of Congress Control Number: 2014943158

Cover and Interior Design by Tom Heffron
Copyedit by John Koharski and Karen O'Brien
Proofread by Leah Noel and Karen O'Brien

On the endpapers: The 1420 owned by Richard Wakeman. *Lee Klancher*
On the back endpapers: Poem by IH project manager Gary Wells about the original Axial-Flow combine. *Case IH*
On the frontispiece: This battered badge has lived on the front of Andrew Tucker's corn picker for five decades—and is still hard at work. *Lee Klancher*
On the dedication page: The first Axial-Flow combine, a 1460, owned by Matt Frey. *Lee Klancher*
On the title page: 1981 International 1420 owned by Richard Wakeman. *Lee Klancher*
On the contents page: A line of combines at work. *Wisconsin Historical Society #114677*

OCTANE PRESS
octanepress.com

Printed in China

Dedication

To Don Murray and Elof Karlsson

Contents

McCORMICK-DEERING

CASE

Preface

THE COMBINE FAMILY

By Gerry Salzman

I grew up on a livestock and grain farm near Ashton, Illinois. My dad relied on farm equipment to feed our family, and today my nephew and brother-in-law do the same on our family farm.

When my dad upgraded to an Axial-Flow combine, the machine made it easier to harvest higher quality grain and more of it. I saw that firsthand and am proud to have played a role in bringing that machine and its successors to farm families around the country.

Axial-Flow combines dramatically impacted these people's lives—I saw that while visiting farmers and dealers during my 42 years with IH and Case IH. I grew from a local territory manager to a role in which I helped guide product development globally for red harvesting equipment. During that time, our customers and business colleagues became my friends.

It has been very enjoyable working on this book in conjunction with Lee Klancher, and I'm pleased that the result is a collection of stories from the people who live and breathe combines. Families from all over the world who count on the efficiency and reliability to harvest their crops and, ultimately, feed their people. The book's narrative also speaks to the dealers and company families worked so hard to bring this innovation to market.

The creation of the Axial-Flow combine as we know it today was not done easily. The original required a massive leap of faith on the part of first a core group of people who believed in the concept and kept it alive in the 1950s and 1960s, and then the entire company that came together to transform the industry in the 1970s.

Since that time, growth and evolution have required risk, courage, sweat, and tears before success, sometimes more of the former than the latter. I'm proud to have played a small role in the successes we achieved along with some great business colleagues. I have really enjoyed working with Lee to present the highs and lows we experienced in creating the Axial-Flow combine.

Creating a good combine isn't easy or always pretty, but the results are worth the effort. Red combines have helped a lot of good people I care about make a living. It has been an honor to have worked so closely for so long with these people, and I am thrilled that their story is being told in this book.

It's a great story, full of drama and hope and suspense and struggle. It is my hope that you enjoy reading it as much as I did living it.

Salzman Family

◀ Gerry Salzman with Ralph, his father, and Adam, his son, in the early 1980s.

Salzman Family

McCORMICK
MADE IN U.S.A. INTERNATIONAL HARVESTER COMPANY CHICAGO
INTERNATIONAL HARVESTER
BE CAREFUL

Introduction

SEEDS OF INNOVATION

By Lee Klancher

Some of the greatest stories of innovations come from motorcyclists stranded in remote corners of the world. People at the edge of the planet repair worn-out pistons with beer cans, resurface valves with pieces of rubber hose, and patch holed crankcases with chewing gum and duct tape.

The lesson that comes from this, I believe, is that when human beings are put in tough situations, they will solve the problem with what they have at hand.

Motorcycle adventure riders are weekend warriors and amateurs when compared to farmers. Farmers spend their lives in a metaphorical wilderness and are constantly confronted with problems that need to be solved in a few hours or minutes in order for them feed their families.

Farmers have welders. Farmers need mechanical aptitude to get through the day. And they often spend hours on a combine or a tractor doing work that allows some time to ponder things a bit and come up with a new solution.

Given these parameters, the fact that farmers are natural-born innovators is no big surprise.

Farm innovation happens in back pastures and on long days cultivating, planting, and combining. Right now, somewhere, a farmer is cooking up a new method to plant crops or otherwise transform and improve his world.

The rotary combine was an idea that sprang from the farm, one that many people kicked around for years. Separating grain required massive amounts of time. As the industrial revolution came on, people around the globe focused on finding more efficient ways to perform routine tasks.

The concept was an old one by the time International Harvester engineers started working on it, but execution was expensive and more difficult than anyone anticipated. Developing the Axial-Flow combine would require millions of man hours and billions of dollars.

Such investment made no sense from a practical standpoint. When the bean counters ran the numbers, the result was the same: too much cost, too little return.

Thankfully, an assortment of people with vision, persistence, and a reckless lack of respect for bean counters decided that the rotary combine concept needed to exist.

For more than a decade, rules were bent, budgets were siphoned, and midnight oil was burned in order to make the rotary combine a reality. And once the machine was built and on the market, another four decades were spent refining and perfecting that technology.

All that work was done by a group who grew up on the farm, living and breathing in a fertile environment that tested both the combines they built and the mettle of the men and women who grew up laying on the ground fixing broken manure spreader chains in 20-below-zero weather and thinking, *"Darnit, there has to be a better way."*

This is the story of the people who created a transformative piece of equipment that has helped shape the farm as we know it today.

Working Man

◀ Daniel J. Tordai and his McCormick 141 in action.

Lee Klancher

Prologue

HARVESTING AMERICA

By Lee Klancher

"The production and harvesting of the grains of the world is basic to our survival. The developments in other disciplines have received more attention but none is more significant."
—Don Murray, unpublished memoirs

Matthew Gregory was born in Vermont in 1802. In 1815, severe frosts destroyed his family's crops and his only sister died. His father—who was 50 years old at the time—decided to seek his fortune elsewhere. In February 1816, the family loaded what they could carry into wooden-shod sleds hauled by four ox teams and headed west.

"We had good teams," Gregory wrote, "but we had a tedious journey." The rough ground required the sleds to be reshod each day, and they had to be fitted with wheels for part of the journey. They arrived in Shelby, New York, in March and spent a few weeks procuring 100 acres of land, for which they paid $100. By early May, the family had built a log cabin and was focused on clearing land.

The year 1816 was a notoriously cold one, with snow falling in all but two months. The Gregory family planted peaches and corn, neither of which produced much of a crop.

A year later, the family planted their first crop of wheat. "This crop, though sowed among roots and stumps of trees, produced a yield of 30 to 50 bushels per acre," Gregory wrote. Harvesting 30 bushels of wheat at that time required roughly 100 hours of labor. The nearest grist mill was nearly 50 miles away, and selling the unprocessed wheat proved nigh impossible. Gregory wrote, "In many cases, green wheat was boiled whole and eaten with milk."

Gregory survived those rugged early days and spent the balance of his life farming his father's piece of ground near Shelby, New York. His father lived to the ripe old age of 72, serving as the local preacher.

In the 1840s, Gregory was able to harvest ten acres of wheat a day thanks to the invention of the reaper. At that time, 70 percent of the American population were subsistence

Early Farm Family

◀ Farming at the turn of the century was back-breaking labor, and it was done by a large portion of America. In 1870, more than half of the American population were farmers. *Wisconsin Historical Society #117321* farmers. The impact of the reaper on America far outstripped anything we have experienced in our lifetimes—an agrarian way of life gave way to the urban world in which we live today. We grumble about adapting to smart phones or email communication; compare that to giving up a hard-working farming life to work in a factory or office in the dirty, dangerous cities of the turn of the century.

The man who took credit for invention of the reaper, Cyrus McCormick, built an empire selling the machines. His company grew to distribute reapers and other agricultural equipment around the globe.

By 1902, his company was merged with most of his competitors to become the International Harvester Company, the world's largest manufacturer of agricultural equipment. The company controlled more than 70 percent of one of the largest businesses on Earth at the time. The McCormick family would marry into other families of privilege, including the Rockefellers.

Today, we consider Microsoft and Apple nascent giants who have ridden a wave of changing technology. Neither has as large a market, or as tight a death grip on it, as the International Harvester Company had on America in the early 1900s.

Industry was transforming the world at the time, and the International Harvester Company used market power and deep pockets to produce equipment that transformed agriculture. Without their industry's efforts, Steve Jobs and Bill Gates might have been nothing more than frustrated bookkeepers who spent their evenings in the garage constructing elaborate abacuses.

Source: Matthew Gregory's story from *Pioneer History of Orleans County* (1871) by Arad Thomas, Orleans American Steam Press Print.

Reaper Invention

▲ While more than 50 reaper designs were built and tested between 1786 and 1840, Virginian Cyrus Hall McCormick was the first man to market and sell the reaper in volume. This famous painting shows McCormick's first demonstration of his machine in 1831. *Wisconsin Historical Society #4993*

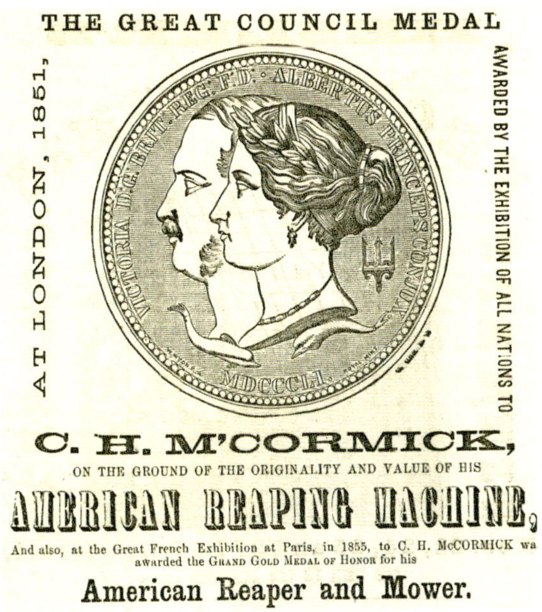

The Great Exhibition Handbill

◀ The Great Exhibition in London was a massive fair held in 1851 to showcase the world's best technology. McCormick was selected to represent the American farm machinery industry. These fairs were lavish events attended by the time's most prominent intellectuals, industry leaders, and royalty. Most of the wares were housed in elaborate temporary buildings constructed for the event, and McCormick's display was woefully unimpressive by comparison. On opening day, the *London Times* disparaged the McCormick display and wrote that the reaper appeared to be "a cross between a flying machine, a wheelbarrow, and an Astley chariot." The reaper redeemed itself with a great performance in the field, trouncing the competing machines. The world applauded, and even the *Times* changed its tune, writing, "[T]he McCormick Reaper was doubtless worth more than the cost of the entire Exhibition."

Wisconsin Historical Society #35012

Westward Expansion

◀ As America's pioneers explored the vast frontier of the West, demand for reapers grew. The McCormick Harvesting Company expanded and built factories in the Midwest to reduce shipping costs to the thousands of farmers taking advantage of fertile new farmlands.

Wisconsin Historical Society

Civil War

▲ The American population was about 31 million people when the Civil War broke out in 1861. The conflict sent 3 million men—nearly 10 percent of the population and a huge portion of the American workforce—into battle. By the time the bloody war ended in 1865, more than 620,000 men had died. The resulting shortage of men available to work at home created an unprecedented demand for improvements in agricultural technology. *Wisconsin Historical Society #41705*

Reaper Wars

▲ McCormick was one of the largest manufacturers, but the reaper was a hot item and hundreds of companies joined the fray. To sell the machines, manufacturers held field demonstrations that pitted the reapers head-to-head. With orders on the line, salespeople were famously willing to use any technique necessary to land a sale—including slander, sabotage, and even fisticuffs.

Wisconsin Historical Society #114305

Obed Hussey

▲ McCormick's biggest competitor was Obed Hussey. The inventor from Maine patented his reaper in Ohio in 1833, a year before McCormick. The two battled tooth and nail over the rights. The market eventually determined a winner, and Hussey was driven out of business and forced to sell his patent to McCormick in 1858. Two years later, Hussey fell while attempting to board a train and was killed.

Wisconsin Historical Society #24855

1883 Field Trial Advertisement

▶ When a manufacturer took home the spoils, it placed advertisements to take best advantage of its feats.

Wisconsin Historical Society #32310

Agricultural Industry Growth

▲ By 1887, McCormick had moved the center of his operation to Chicago, Illinois, so that his production was closer to his customers. His company had plants all over the world, (including McCormick Reaper Works), and had grown to become one of the world's largest agricultural equipment manufacturers. *Wisconsin Historical Society #39554*

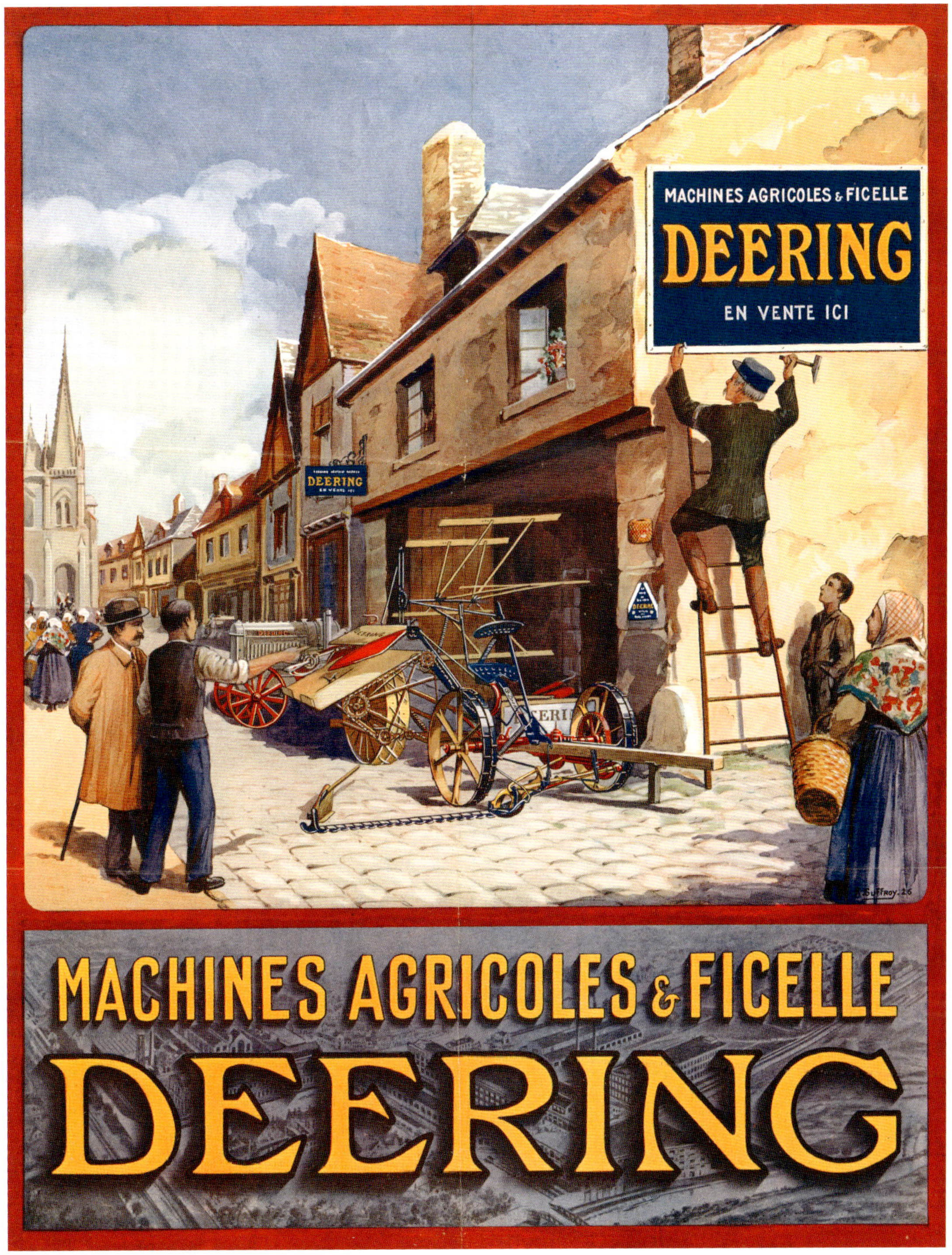

The Reaper Boom

▲ The reaper revolutionized the farm and spread across the globe. This image shows a new Deering dealership in France. Deering was a direct competitor to McCormick. In 1902, McCormick, Deering, and several other companies merged to form the International Harvester Company (IHC).

Wisconsin Historical Society #4322

The World According to IHC

▲ When IHC formed in 1902, its purpose was to dominate the market. Part of that strategy was selling machines across the globe.

Wisconsin Historical Society

The Rise of Power Farming

◀ The invention of large steam and gas tractors allowed farmers to thresh grain, plow open ground, and power machines such as this grain thresher. Increased productivity meant fewer workers were needed to feed the world's rapidly growing population. By 1910, farmers made up only 31 percent of the American population. That trend would continue, particularly once crop-harvesting equipment became more efficient and portable. *Case IH*

1884 McCormick Brochure

▲ McCormick offered a line of harvesting equipment in the late 1800s. This image shows equipment used by the ancient Gauls and a grain binder. *Wisconsin Historical Society #4396*

McCormick Grain Binder in 1899

▲ Iowa farmer harvesting grain with a McCormick grain binder built in 1883. Original caption reads: "The frame was made entirely of wood, iron being used only for gears, braces, etc. McCormick's first twine binder was built in 1881—fifty years after his invention of the reaper. After that time, improvements consisted chiefly in the use of better-grade materials to provide lighter draft and greater durability."

Wisconsin Historical Society #22888

McCormick Improved Grain Binder

▶ By the 1910s, McCormick was a well-established brand. This 1917 color advertisement pushes the deep history of the line that started with Cyrus McCormick inventing the reaper. *Wisconsin Historical Society #50117*

Mogul and Harvester-Thresher

▶ As IH began to develop farm tractors, the equipment naturally became larger to match the increased pulling power of a tractor. This circa-1912 photograph marks the harvester-thresher as an experimental model. *Wisconsin Historical Society #49199*

Mogul 25-hp with Grain Binders

▼ The appeal of harvesting a lot of ground with a big lineup of equipment appeared to begin as soon as machines were created. This image from 1912 shows a Mogul 25-hp tractor pulling three McCormick grain binders. *Wisconsin Historical Society #73408*

Jerome Increase Case

▲ J. I. Case started manufacturing farm machinery in a small shop in Racine, Wisconsin, in the 1840s. He grew that business to supply machinery in the central Midwest and founded the J. I. Case company in 1863. J. I. Case grew to be one of the world's largest manufacturers of agricultural equipment. *Case IH*

J. I. CASE THRESHING MACHINE COMPANY

The legacy of modern red combines includes the work of Jerome Increase Case. In 1844, he founded the J. I. Case Threshing Machine Company, and that company grew to be one of America's largest manufacturers of steam engines. The company produced self-propelled portable steam engines, steam-powered tractors, threshing machines, and other harvesting equipment.

The company introduced the first all-steel threshing machine in 1904, and sold its first gas tractor that year. The company had a strong presence in Europe at that time.

In 1928, the company would become the J. I. Case Company, and under that banner built tractors and, eventually, combines. The company was one of the ten largest agricultural manufacturers in America during the twentieth century.

The distinctive image of Old Abe was a recognized portion of the J. I. Case logo. Old Abe was a real eagle and the mascot of Company C of the Eighth Wisconsin Regiment. Jerome Increase Case encountered Old Abe in 1861. The eagle was in a parade with Company C in Eau Claire, Wisconsin. Case inquired about the bird's story and learned Old Abe had been through 38 battles with Company C. He became determined to adopt the bird as his company's logo once the war was over, and in 1865 Old Abe became the trademark of the J. I. Case company.

The name of the company changed to simply Case under Tenneco ownership in 1970. The brand merged with International Harvester in 1984, when Case IH formed.

J. I. Case Steam Engine

▶ One hallmark of the J. I. Case company was its steam tractors. In the late nineteenth century, J. I. Case was one of the world's largest suppliers of steam engines. *Case IH*

French Advertisement

▲ Like the International Harvester Company, J. I. Case exported its products to locations around the world. Jerome Increase Case owned several ships that worked the Great Lakes—they may have been part of the inspiration for this advertisement. *Case IH*

J. I. Case Threshers

▲ While the company was best known for steam tractors, the J. I. Case threshers were also well established and popular machines. *Case IH*

Chapter One

1915–1941 THE EARLY DAYS OF COMBINES

By Lee Klancher

"As respects this ingenious machine, it remains only to say that it harvests, cleans, and *bags* from twenty to thirty acres of heavy wheat, in the course of a single summer's day!"
—James Fenimore Cooper, *The Oak-Openings* (1848), writing about Hiram Moore's combine

While Cyrus McCormick was building his reaper empire, two men in Michigan were building the next transformative piece of agricultural technology. In June 1836, Hiram Moore and John Hascall of Kalmazoo patented a machine pulled by 14 horses that harvested, threshed, cleaned, and bagged grain in one pass. By summer 1838, the men were able to bag 1,100 bushels of wheat in one day. They sold several of their combines for $500 each, at least one of which worked for 20 years.

The machine was the combine. Their idea spread to California and by 1888 roughly 500 of them were working on the West Coast. Many of these were very large machines that had a 40-foot cut and were hauled by large horse or mule teams. The technology was cutting-edge, as was the $500 price—an entire farm could be purchased at the time for $50 and an annual budget of $200 or so was not uncommon. Combine purchases were suitable only for large, well-funded operations that worked 300 acres or more.

In 1845, Australians had developed a smaller machine, the stripper-harvester, that stripped grain off standing stalks. The machine worked well in light crops and was reasonably light and economical.

By 1887, the big combines were popular out West, and the stripper harvesters from Australia had migrated elsewhere. Similar developments were being made around the world.

While blacksmiths, visionaries, and schemers developed new ways to improve productivity on the farm, Axel Jarlson was fighting to stay alive with his family in Sweden. The winter of 1891 produced deep snow, six months worth, and the only way the farmer's family could survive was to make wood carvings, cabinets, and mittens and sell them in town, ten miles from their farm.

Bountiful Harvest

◀ The International Harvester Company was the world's largest agricultural machinery company when it began manufacturing combines in 1915. *Wisconsin Historical Society #9258*

World War I

▶ World War I began in 1914. The United States was in a recession in 1913, but exports for the war effort pulled International Harvester Company back into the black. When the United States entered the war in 1917, it was ill-prepared to deal with the increased demand. By 1918, however, the country had created 5,000 new federal agencies that employed more than half a million people. The new agencies transformed the United States into a wartime production machine.

Wisconsin Historical Society #3548

"Swedish people who have money hold on to it very tightly," Jarlson wrote, "and often we took things to market and then had to bring them home again, for no one would buy."

Jarlson's parents were sick the entire winter and all they had to eat was black bread and gruel. They had to sell the cows, and had no milk or cheese.

Jarlson's older brother, Gustaf, left for America that winter. Their uncle paid $90—a vast sum—for Gustaf's ticket to Minnesota. In 1899, Gustaf sent money for Jarlson and his sister to come to join him. Jarlson worked for his brother in Minneapolis and made $16 a month. He saved it until he had $304, then bought a farm for $150 and spent another $55.42 on supplies. He worked all winter clearing land and cutting wood, and planted 12 acres of land in April.

His 67-bushel wheat crop that year netted him $46.50. He augmented that by selling firewood for $1.25 per cord, his sister's home-canned preserves, and other crops to gross $530. His expenses were $552.17, meaning he lost $22.17 during the course of the year. But the thrifty Swede had cash on hand and survived into 1902, when he grossed more than $1,200 and netted enough profit to afford to travel back to Sweden to celebrate Christmas with 600 of his fellow countrypeople.

He carried with him photographs of American women, one of whom had married his brother, Gustaf. "The Swedes who live in America like the old country girls," he wrote, "because they know how to save money."

While Jarlson was carefully managing his Minnesota farm, the International Harvester Company was busy finding ways to provide people like him an efficient machine that would improve his profitability and persuade him to part with his hard-earned cash.

In 1904, the International Harvester Company (IHC) entered the stripper-harvester market. The company distributed a number of thresher lines as well. From 1910 to 1912, IHC's engineering team worked to develop a light harvester-thresher. IHC purchased the rights to a push-type machine from Thomas Dugan in 1913, but the machine proved unsuccessful. In 1913, the company developed its own harvester-thresher. That early machine was a prairie-type combine, meaning it was lighter, more economical, and designed for use on small farms. It was refined and tested and, in 1915, put into production as the Deering No. 1 and McCormick No. 1.

With such a machine, men like Jarlson could harvest 50 bushels a day. The increased productivity transformed his operation and could allow him to farm 100 or more acres each

Wartime Production

▶ Between 1914 and 1917, American industrial production increased 32 percent. Gross farm income also skyrocketed, increasing 230 percent from 1914 to 1919, as did the demand for agricultural equipment. This photograph shows IHC's Milwaukee Works in 1914. *Wisconsin Historical Society #7601*

year, rather than just 12. Perhaps the thrifty Jarlson purchased a combine and was able to find one of those country girls he liked. One could further speculate that his operation might have become profitable enough that his first concern wasn't his potential wife's ability to save money.

While it's doubtful the harvester-thresher could change the character of a thrifty Swede, Jarlson's children would grow up in a world that reaped the benefits of his increased productivity.

Source: Axel Jarlson's story from *Undistinguished Americans* (1906) by Hamilton Holt, James Pott & Company.

Wartime Call for Technology

◀ After World War I, the United States needed to step up farm production. This advertising illustration, which appeared on the exterior of a mailer advertising the Illinois Tractor Company, promoted the use of tractors and engine power over horses and horse-drawn implements. "The Problem of Peace" refers to challenges in the wake of World War I. *Wisconsin Historical Society #97153*

McCormick No. 1 Harvester-Thresher

▲ Developed from 1910 to 1913, the first McCormick harvester-thresher was a lightweight machine powered by the ground wheels. These lightweight machines were referred to as "prairie-type" harvester-threshers. The No. 1 was a limited production model in 1913, and in 1915 IH introduced the machine for regular production. The straw was carried from the machine by a vibrating carrier extending at right angles from the body of the machine. Note the bagging type of grain box, which could be emptied while the machine was in operation. By March 1919, IHC literature referred to its line of harvester-threshers as "combines." The name "thresher-harvester" was not officially struck until April 14, 1959, when the product identification committee decided that the machines would be referred to as "combines" in all company documentation.

Wisconsin Historical Society #114632

1913 McCormick Experimental Harvester-Thresher

▲ International Harvester began experiments with its own harvester-threshers in 1913. This unit shown in April 1913 was one of IHC's first experimental harvester-threshers. This is the pull-type variety—the company also tested a push-type version of this machine. *Wisconsin Historical Society #7647*

Deering No. 1 Harvester-Thresher

▲ The Deering No. 1 was developed in conjunction with the McCormick model. The first Deering experimental harvester-thresher was built at Deering Works in 1913. Its design was essentially sound, and the No. 1 became the basis of subsequent IHC harvester-thresher construction for many years. Deering harvester-thresher Nos. 2 and 3, and McCormick Nos. 2 and 6 were patterned after it. The Deering No. 1 differed from the later models in that it employed a side shake to the riddles and had a "U" shape main axle. The No. 1 had a cutting width of 9 feet. This image was taken July 9, 1915, on a factory or warehouse floor. *Wisconsin Historical Society #45642*

McCormick No. 2

▲ First built in 1916, the No. 2 was the first McCormick-built machine to use straw walker–type separating mechanics. The No. 2 had a double cleaning shoe with two cleaning fans and was chain-driven from the main wheel. The cutting width was 9 feet. This machine was sold domestically as well as abroad. This image was taken March 19, 1917. *Wisconsin Historical Society #114634*

Deering No. 2

▲ This is an engineering photograph of a Deering No. 2 harvester-thresher equipped with bagging-type grain tank. Like the Deering No. 1, this machine was gear-driven from the main wheel. Its cutting width was 9 feet. A 3-foot platform-extension option offered a total cut of 12 feet for light grain. Image dated April 22, 1916, with a location listed as Chicago. *Wisconsin Historical Society #25451*

Deering No. 3

▲ The Deering No. 3 harvester-thresher as first built in 1923. The gear drive of the earlier machines was replaced with a chain drive. The slated feeder elevator was a distinguishing feature of this machine in 1923. The following year the elevator was replaced by a feed conveyor. Image dated March 12, 1923.

Wisconsin Historical Society #114636

McCormick No. 4

▲ The McCormick No. 4 harvester-thresher as built for Argentina in 1925. This model was originally developed for the domestic trade in 1924 but only a limited number (approximately 100 machines) were put out that year. Thereafter, it was sold only in South America, reaching a peak production of 1,500 units in 1926. Image dated June 14, 1925. *Wisconsin Historical Society #114637*

McCormick-Deering No. 5

▲ The No. 5, along with the No. 4, may be said to mark the transition from the old-style harvester-thresher construction dominated largely by the original Deering design to the more modern and efficient machines of the No. 11 type. Image dated July 15, 1925. *Wisconsin Historical Society #45641*

McCormick No. 4

▲ A McCormick No. 4 combine working an Argentine field in 1926. The No. 4 was the so-called "low feed" type of harvester-thresher and was a forerunner of the No. 10 and 11 machines. Weighing approximately 7,500 pounds, it had a rigid 12-foot platform and an engine mounted above the main axle. The No. 4 was replaced by the No. 10 in 1927. *Wisconsin Historical Society #114638*

McCormick No. 6 Pull-Type

▲ International Harvester exported this 9-foot combine to South America in February 1922. Many of the company's combines were shipped abroad, at times under slightly different monikers: Deering, McCormick, and International were common, but other brand names were used on occasion. *Wisconsin Historical Society #117393*

McCormick-Deering No. 7 Hillside

◀ Farmers harvesting grain with a horse-drawn McCormick-Deering No. 7 Hillside harvester-thresher. The machine incorporated a leveling device on the outer wheel that could be changed at any time as required, while the header was free to pivot to accommodate varying slopes. Image taken September 21, 1926, on the farm of Mr. Joy McGuire in Thorton, Washington. *Wisconsin Historical Society #11009*

McCormick-Deering No. 7 Hillside

▲ The No. 7 Hillside machine weighed 7,500 pounds and was designed to be operated by two people. This image shows the same machine and farm shown in the previous image.

Wisconsin Historical Society #11010

McCormick-Deering No. 8

◀ Men harvesting wheat with a McCormick-Deering Model W-30 tractor and a Model 8 combine equipped with an auxiliary engine. Note that auxiliary engines became increasingly popular over the years and were needed for heavy crops. This tractor and combine were owned by George Todd and were equipped with Firestone tires. Image taken August 9, 1935, in Delphi, Indiana.

Wisconsin Historical Society #7225

McCormick Type C Harvester-Thresher

▶ Built from 1929 until at least 1932, the Type C harvester-thresher was similar to the No. 8, but was modified to supply Agar, Cross and Company with a machine sufficiently different in appearance to be sold as a competitor to the No. 8. The Type C was equipped with oval elevators, McCormick-type wheels, and several other minor modifications. It was also painted differently than the McCormick model. Only 582 machines were built between 1929 and 1932. Image taken December 1932. *Wisconsin Historical Society #114639*

1926 McCormick-Deering No. 9

▲ The No. 9 was a large 16-foot machine and the heaviest combine built by the Harvester Company at the time. Its principal characteristics were a floating-type header platform, drag-type feeder, ball bearing cylinder, four-section straw rack, re-cleaner, and 60-bushel grain tank located on top of the machine. It weighed approximately 10,650 pounds. Image taken September 6, 1926.

Wisconsin Historical Society #114640

McCormick No. 10

▲ A No. 10 combine equipped with 16-foot solid platform. Beginning in 1927, this machine was the accepted heavy-type machine sold by Agar, Cross and Company in Argentina and by other McCormick dealers elsewhere abroad. Image taken March 22, 1927.

Wisconsin Historical Society #114645

McCormick-Deering No. 11

▶ The No. 11 was the updated replacement for the No. 4 model. It was a large machine powered by an auxiliary engine.

Wisconsin Historical Society #117319

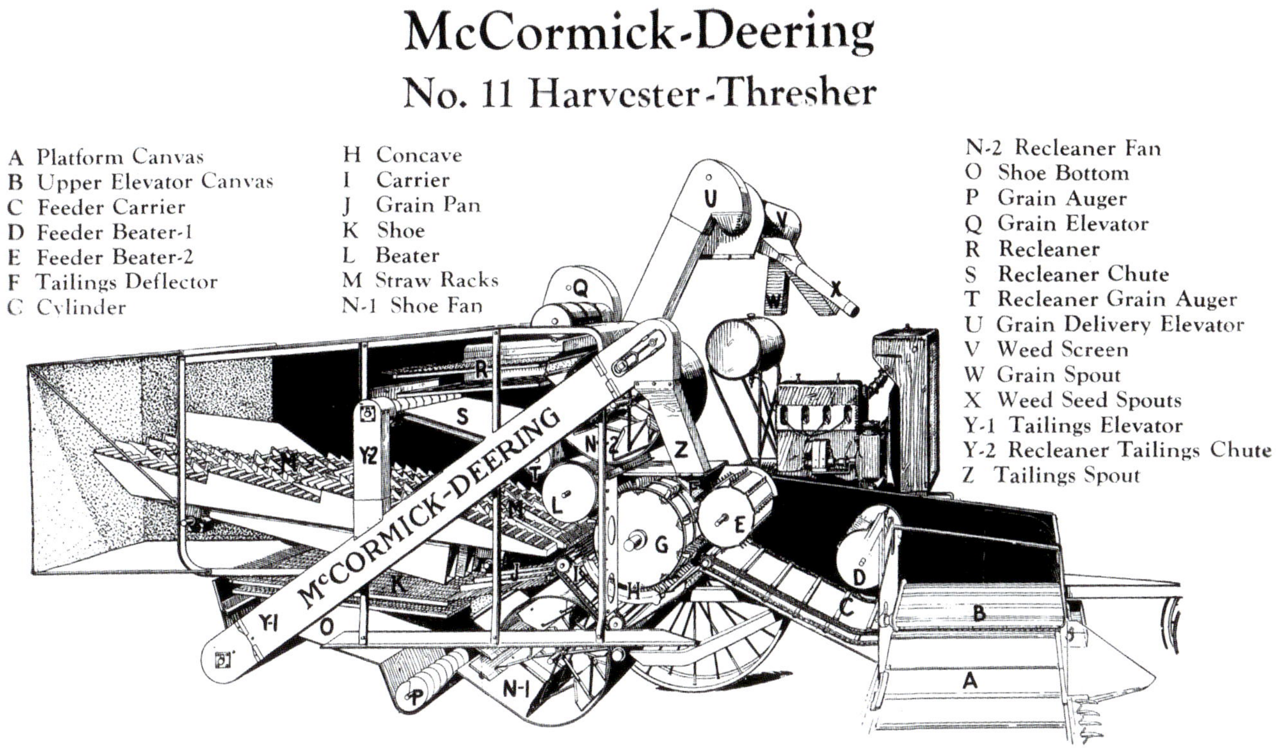

McCormick-Deering No. 11

▲ Louis Bertrand of Oakley, Kansas, and his fleet of 12 McCormick-Deering 15-30s and 12 No. 11 combines. *Wisconsin Historical Society #117323*

McCormick-Deering No. 20

▲ A No. 20 combine equipped with pickup attachment. When this device was used on the No. 20 harvester-thresher, the machine had to be equipped with an auxiliary engine. *Wisconsin Historical Society #114647*

McCormick-Deering Type N Experimental Combine

◀ This shows the Type N experimental harvester-thresher at work. The Type N machine was an enlarged version of the No. 20 harvester-thresher with 10-foot cutting platform, four-section straw rack, and auxiliary engine. The basic design (similar to that of the No. 20) proved so satisfactory that it was later adopted for the Nos. 31 and 41. Image taken during the 1930 harvest season near Hutchison, Kansas.

Wisconsin Historical Society #114649

McCormick-Deering No. 21 Harvester-Thresher

▲ The No. 21 was built expressly for European grain conditions and was equipped with a torpedo divider, grain-type reel, and operator's platform. Image taken in 1932.

Wisconsin Historical Society #114651

McCormick-Deering No. 22 Harvester-Thresher

▲ Three-quarter rear view of a McCormick-Deering No. 22 harvester-thresher combining soybeans on the Purdue University Soils and Crops farm near Lafayette, Indiana, and pulled by an F-20 Tractor. The No. 22 could be equipped with rubber Firestone tires. Image taken February 28, 1928. *Wisconsin Historical Society #78108*

McCormick-Deering No. 22 Harvester-Thresher

▶ The No. 22 was an engine-driven combine with a relatively small 22.75-inch-wide cylinder. Power was provided by an 18-hp engine and the standard platform was 8 feet wide. *Wisconsin Historical Society #117395*

McCormick-Deering No. 22 Stationary Harvester-Thresher

▲ International made several No. 22s—both the pull-behind model and a large stationary thresher shown here. This International 22x38 thresher was owned by Otto Strange and is being driven with his Titan 10-20 tractor. *Wisconsin Historical Society #115110*

McCormick-Deering No. 28 Stationary Harvester-Thresher

▲ This model was similar to the No. 22 stationary, though larger.

Wisconsin Historical Society #117403

McCormick-Deering No. 30

▲ A right-side view of a No. 30 harvester-thresher with header platform removed. This medium-capacity 12-foot harvester-thresher was built in limited quantities in 1930 and 1931. The construction was very similar to the No. 40. Image taken June 14, 1930.

Wisconsin Historical Society #114652

McCormick-Deering No. 31

▶ Farmer Frank Rakow harvesting "gyp" (Egyptian) corn with a McCormick-Deering No. 31-RW West Coast Special combine on a 200-acre island farm leased by W. T. Jarrett from Liberty Farms Company. The delta region formed at the confluence of the San Joaquin and Sacramento rivers in California is broken up into some 40 to 50 large islands up to 20,000 acres in size. The islands, most of wich are owned by corporations, are protected by dikes and are usually lower than the rivers. Since land is very fertile and water plentiful, huge yields prevail. Image taken in Rio Vista, California, in 1935. *Wisconsin Historical Society #25461*

McCormick-Deering No. 31-RW West Coast Special

▲ New 31-RW West Coast special harvester-thresher owned by Frank Rakow of Rio Vista, California, and pulled by an off-brand tractor harvests on a 200-acre island farm leased by W. T. Jarrett from Liberty Farms Company. Shown, left to right, are H. C. Sage, assistant branch manager, general line brand, San Francisco; G. C. Gordon of Gordon, Hansen Company, McCormick-Deering dealers, Rio Vista; and Jarrett and Rakow. *Wisconsin Historical Society #114656*

McCormick-Deering No. 31-RW West Coast Special

▲ New 31-RW West Coast Special harvester-thresher in gyp corn on the 110-acre farm of T. C. Rawlins in Glenn, California. The machine was owned by Guey Jones of Willow, California. *Wisconsin Historical Society #78552*

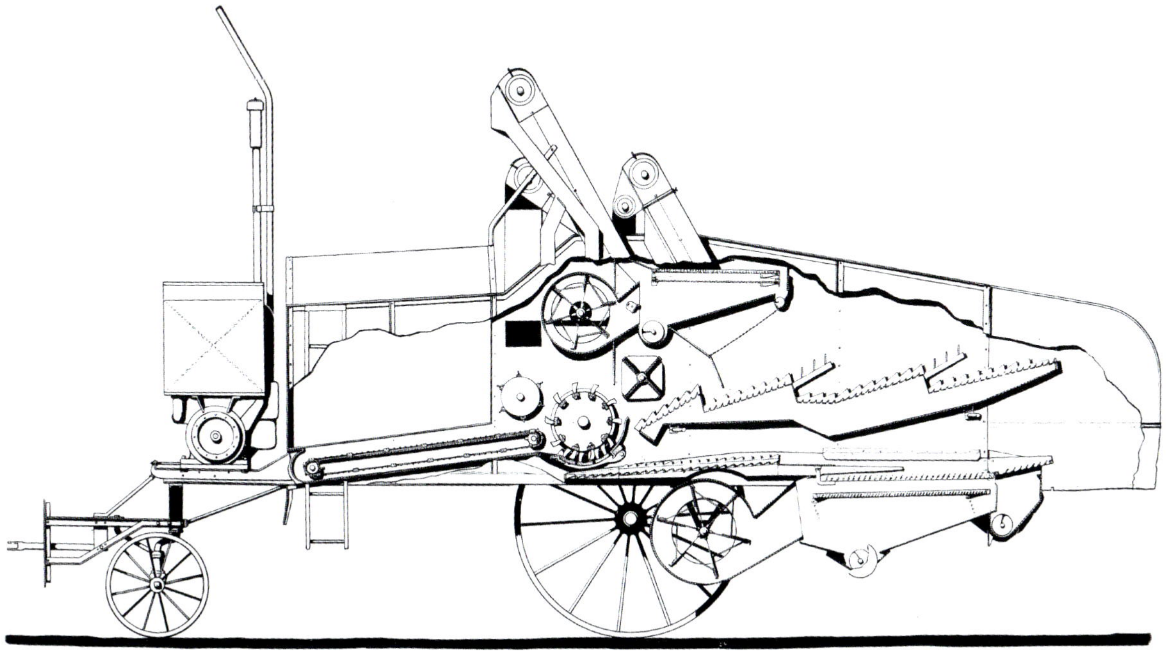

McCormick-Deering No. 31-T

◀ This drawing was for the export version of the No. 31-T and was marked, "Negative sent to Buenos Aires."

Wisconsin Historical Society #114661

McCormick-Deering No. 36 Thresher

▲ The No. 36 was a stationary model that could be configured with an optional trailer attachment that allowed it to be transported. Designed to be powered by a 7- to 14-hp tractor belt pulley or stationary engine, the No. 36 was designed with the small-acre farm in mind. *Wisconsin Historical Society*

McCormick-Deering No. 31-T

▲ The 1935 cover of an advertising brochure for the McCormick-Deering No. 31-T combine.

Wisconsin Historical Society #80928

McCormick-Deering No. 40.

▶ McCormick-Deering 15-30 tractor and No. 40 McCormick-Deering combine, along with an IH A-4 motor truck with a homemade grain box that holds 180 bushels of wheat. *Wisconsin Historical Society #114665*

McCormick-Deering No. 41

▲ Elevated view of a farmer harvesting grain with a McCormick-Deering tractor and No. 41 16-foot combine. The No. 41 featured a 28-inch cylinder and 42-inch separator and was equipped with a re-cleaner and 45-bushel grain tank. At the time of this photograph, only four of these machines had been built (one of them for export). It was expected that the No. 41 would go into full production shortly after this image was taken and as the existing stock of No. 11 combines were depleted. Image taken in 1932 near Quinter, Kansas. *Wisconsin Historical Society #25458*

McCormick-Deering No. 41

▲ The No. 41 was a three-wheel in-line machine equipped with 16-foot folding-type platform, five-section straw rack, re-cleaner, and platform power control. The latter device permitted the operator to raise or lower the platform as desired by pulling one of two ropes conveniently attached to the tractor. Image taken in 1932 near Quinter, Kansas. *Wisconsin Historical Society #25459*

McCormick-Deering No. 51 Combine

▶ Two men harvesting wheat with an International crawler tractor (TracTracTor) and a No. 51 Hillside combine. The second man on back levels the machine. Image taken near Bozeman, Montana, in 1950.

Wisconsin Historical Society #24614

McCormick-Deering No. 60

▲ McCormick-Deering No. 60 pulled by an F-20 Farmall combines wheat on the Purdue University Livestock Farm near Lafayette, Indiana. The No. 60 combine was PTO-driven. Image taken September 30, 1937. *Wisconsin Historical Society #78039*

McCormick-Deering No. 60 Combine

▲ An advertising poster for the McCormick-Deering No. 60 combine touts the machine's efficiency and ease of use. *Wisconsin Historical Society*

McCormick-Deering No. 61 Combine

▼ Unloading wheat from a No. 61 combine on July 11, 1939, on the Purdue University farm near Lafayette, Indiana.

Wisconsin Historical Society #114675

McCormick-Deering No. 42

▲ The No. 42 marked a change in style for International Harvester's combines, with a more rounded look. This No. 42 worked on the farm of Raymond Urfer and his father, Charles, who farmed 160 acres of flax and soybeans. They purchased a No. 42 to help with harvest so they would have time to care for their dairy herd. Their farm was purchased when Raymond was five or six. Charles used three Farmalls and had just added the Farmall H when Raymond married and took on another farm. Image taken near Des Moines, Iowa.

Wisconsin Historical Society #114668

McCormick-Deering No. 52R

◀ One of the most popular International Harvester pull-type combines was the No. 52R. The machine had a 5-foot cut and an 18-bushel grain tank and was produced during the same period as the larger No. 62. More than 36,000 No. 52Rs were produced between 1940 and 1951. This Farmall H and No. 52R are owned by Daniel J. Tordai. *Lee Klancher*

McCormick-Deering No. 62

▼ Like the No. 60 and 61, the No. 62 had a 28-inch-wide cylinder and a 6-foot platform but was otherwise a completely new model that featured the styled sheet metal and bright paint of the No. 42.

Wisconsin Historical Society #114663

J. I. CASE COMPANY 1915–1941

In 1923, the first Case combine was built by the J. I. Case Threshing Machine Company. The company introduced a line of small gasoline tractors that were painted gray. J. I. Case stopped producing steam tractors in 1927. It had built more than 30,000 steam tractors before ceasing production.

In 1928, the company name changed to the J. I. Case Company. The company had a strong presence around the world, including sales outlets in Sweden, Australia, and Mexico.

Interestingly, Case continued to sell its threshing machines long after the small combine was introduced. Well into the 1930s, crews traveled America with threshing machines to harvest farms. The machines required 20 to 30 people to operate a full-scale threshing. Quite often, a group of families would work together to thresh their grain with the contracted help.

The power of bright colors became understood in the 1930s, when many of the tractor manufacturers switched from drab blues and grays to brighter, more eye-catching hues. J. I. Case followed suit in 1939, painting its agricultural equipment Flambeau Red. The color was a shade of orange, and four decades would pass before a merger created Case IH in 1985.

Horse-Drawn Combine

▼ A Case hillside machine is drawn by a team of horses. Case provided equipment to hook its combines to horse teams well into the 1930s.

Case IH

1936 Model C

◀ The Case Model C was introduced in 1936 and eventually made available in 8-, 10-, and 12-foot styles. *Case IH*

1936 Model Q

◀ A larger model Case combine. *Case IH*

Chapter Two

1942–1961 THE RISE OF SELF-PROPELLED MACHINES

By Lee Klancher

"The war has made the farmer almost the most important person in the country, and farming has become as essential a war-time business as the manufacturer of planes, tanks, guns and ammunition."

—Editor of an Oklahoma newspaper, quoted by R. Douglas Hurt in *The Great Plains During World War II*

When World War II broke out in Europe, the U.S. government began looking to the farmer to provide additional crops to help supply war-depleted countries overseas. U.S. Secretary of Agriculture Claude R. Wickard believed that American farmers would need to feed 10 million Brits and initiated a nationwide program, "Food for Defense," imploring farmers to step up output. In Kansas, federal and state officials took time to meet with farmers to encourage increased production.

Farmers reacted cautiously at first. Many Americans believed the war in Europe would end quickly, and farmers who could recall the hard times of the Great Depression were reluctant to commit large resources and end up with low prices and surpluses.

That all changed after December 7, 1941. After the Japanese attacked Pearl Harbor, the farm mobilized as quickly and as forcefully as the American military. Prices increased quickly, and farm income nearly tripled between 1940 and 1945.

Which isn't to say these were easy times.

The government strictly regulated farm equipment production, and many farm equipment manufacturers expended their efforts building tanks, guns, and other war machines. They also faced tightly controlled raw material quotas—obtaining steel, for example, was very difficult.

Despite the fact that American farmers of the time could afford and desperately needed new equipment, very few new machines were available. The prices of used equipment soared, and farmers were forced to use old, worn-out equipment.

Labor was also critically short. More than 15 million people were called into the military—roughly 20 percent of the American workforce—and the pay rate on American farms was lower than in the cities. Getting enough help was a critical issue.

A New Breed

◀ The self-propelled combine was first patented in 1887 and became popular in the 1940s. The machine offered better fuel consumption and was more efficient than pull-behind units. The first IHC version was the No. 123-SP, introduced in 1942. This is an early experimental self-propelled combine at work. *Wisconsin Historical Society #114959*

The solution came from multiple sources. Young boys were trained to work with their brothers and fathers. Migrant labor proved to be a boon to the farm. In 1942, Congress even authorized military deferments for agricultural workers.

Farm women worked on the farm as never before. While Dawn Dyer's husband was training with the Army Air Corps, she ran their farm in Sprague, Washington, with aging equipment and insufficient help. "I'm going to break down and cry pretty soon if we have any more breakdowns during harvest," she wrote her husband in a letter. "This morning they worked on the old Dodge truck for several hours and that held things up. Then this afternoon they broke the header on the combine and ran out of gas."

The Adolf Eifert family near Waukomis, Oklahoma, harvested 300 acres of grain without any additional help.

World War II

▲ As with the Civil War and World War I, the arrival of World War II increased demand for agricultural equipment. With much of the American workforce sent overseas fighting, the need for machines that could maximize productivity increased. This poster was created in 1942. *Wisconsin Historical Society #27681*

War Machines

◀ International Harvester Company supplied equipment to the U.S. military during World War II, including these TD-18 TracTracTor crawlers, shown on Ryukyu Island in Okinawa with Marine Private First Class Gilbert E. Bailey. *Wisconsin Historical Society #64124*

Diversification

◀ International looked to broaden its horizons in the 1940s and early 1950s, investing in construction equipment, home appliances, and a line of commercial trucks. The construction equipment and home appliance ventures proved costly failures. The lack of research dollars funneled into the agricultural division also proved a mistake. It's interesting to note that while this was happening, the combine engineers were developing the Axial-Flow technology in secret without an official budget or backing. *Wisconsin Historical Society #28201*

The *Harold F. McCormick*

▲ The zenith of International Harvester's market dominance came in the late 1940s and early 1950s. The company was the largest agricultural equipment firm by a comfortable margin, and farmers bought power equipment in unprecedented numbers. The company splurged on a few lavish items, such as its first corporate aircraft, which was a converted B-23 light bomber. *Wisconsin Historical Society #47187*

Bon Voyage

▲ Pilot Bill Dotter waves from the Harvester corporate plane. *Wisconsin Historical Society #47190*

"I really didn't even try to find extra hired hands this harvest," Eifert said. "Everybody was hunting for them, and they just weren't to be had. I knew that, and so we just decided to do the work ourselves."

Eifert enlisted the help of his 19-year-old daughter, Helen, who was a student at A&M, and his 12-year-old son, Donald.

Even so, the harvest was one of the family's worst in the 21 years they had farmed the land, yielding only six bushels per acre.

At the International Harvester Company, the self-propelled combine was developed and ready for introduction by 1941. With such a machine, 100 bushels of wheat could be harvested with only 34 hours of labor (compared to 233 hours of labor 100 years earlier). The self-propelled combine offered the kind of productivity necessary to empower families like the Eiferts stay on the farm.

Despite war shortages, the combine came to the farm en masse in the 1940s. More than half a million combines arrived on the farm between 1940 and 1950—more than doubling the entire fleet.

While the number of farmers in America declined, the productivity of those who remained exploded. One key piece of equipment in this puzzle was the self-propelled combine.

Source: "Men at War, Wheat Harvest Story of Women in the Fields," *Daily Oklahoman*, June 27, 1943.

Wartime Combine Production

▶ The self-propelled combines were built at East Moline Works beginning in 1942; production continued until the war shut it down. In 1944, the government asked IH to build No. 123-SPs in order to conserve manpower on American farms. The assembly line ran year-round, and on March 17, 1944, turned out the 10,500th No. 123 combine. This image was taken at East Moline Works in 1944. *Wisconsin Historical Society #7722*

The Self-Propelled Combine Arrives

▲ The big news for 1942 was the release of International's No. 123-SP (self-propelled) combine. The model weighed 7,200 pounds and featured a 12-foot cutter bar and an IH six-cylinder engine. This image shows a machine out on the road with the combine caravan in 1943. This image was taken in a soybean field on the 430-acre Glenn Young farm near Rio, Illinois (7 miles north of Galesburg). What are most likely early production model combines are shown, both a 12-foot-cut unit as well as the new No. 62 combine. Fred De Blieck, engineer in charge of the caravan, is shown at right talking to his tractor drivers and mechanics. Shown are, first row (left to right), H. O. Herrstrom, mechanic; John W. Griffith, mechanic; Marion Griffith, tractor driver; and James L. Amy, tractor driver; and, top row, Ray Griffith and Homer Willard, tractor drivers. *Wisconsin Historical Society #115002*

Fowler McCormick Visits the Caravan

◀ Harvester CEO Fowler McCormick in the field visiting with Mr. Quinby, a field engineer, during the Combine Caravan. McCormick led the company during the 1950s and pushed Harvester to increase its presence in the construction market and to create consumer appliances. These failed diversification programs ultimately saddled the company with a massive debt load.

Wisconsin Historical Society #117285

Combine Caravan

▼ Starting in the early 1940s, International Harvester sent a fleet of experimental combines around the country to be field-tested. The company dubbed the outing the "combine caravan." The caravan included a trailer with enough tools so test engineers could fabricate experimental parts on the spot, and test engineers spent hours working with customers who tested the machines on their farms.

Wisconsin Historical Society #141933

1944 Prototype No. 123-SP

▶ The self-propelled No. 123-SP was sent all over the world. This experimental Deering No. 123 self-propelled combine shown at work in Argentina is equipped with a re-cleaner (extreme left) and grain classifier (back of operator). Image dated 1944. *Wisconsin Historical Society #78542*

1943 No. 123-SP Rice Harvester

▶ In 1943, International Harvester announced the 123-SP self-propelled, one-man, one-engine, 12-foot combine was being equipped with deep-lugged rubber tires and other special equipment for harvesting rice. Its 13.50-28 tires allowed it enough traction to harvest in the wet fields and deep ditches typical of a rice harvest. Special bar and wire-grate concaves were used, and shields were added at the ends of the grain platform, as well as the rear, to prevent loss of grain. Also special for rice harvesting were hook-tooth retarders that held back the material while the stripper beater leveled out so that the rice entered the cylinder and concave in an even stream over their entire width. *Wisconsin Historical Society #15050*

1944 McCormick-Deering No. 123

▶ Two women and a man with a McCormick-Deering 123-SP combine and an International truck. The 123-SP was Harvester's first commercially produced self-propelled combine. Image dated 1944 in Montana. *Wisconsin Historical Society #85076*

McCormick-Deering No. 125-SP

▶ The self-propelled No. 123 was developed fairly quickly after World War II and was the basis for several model changes as new features were added. This image shows a Model No. 125-SP, which was produced in 1948 and 1949. *Wisconsin Historical Society #117286*

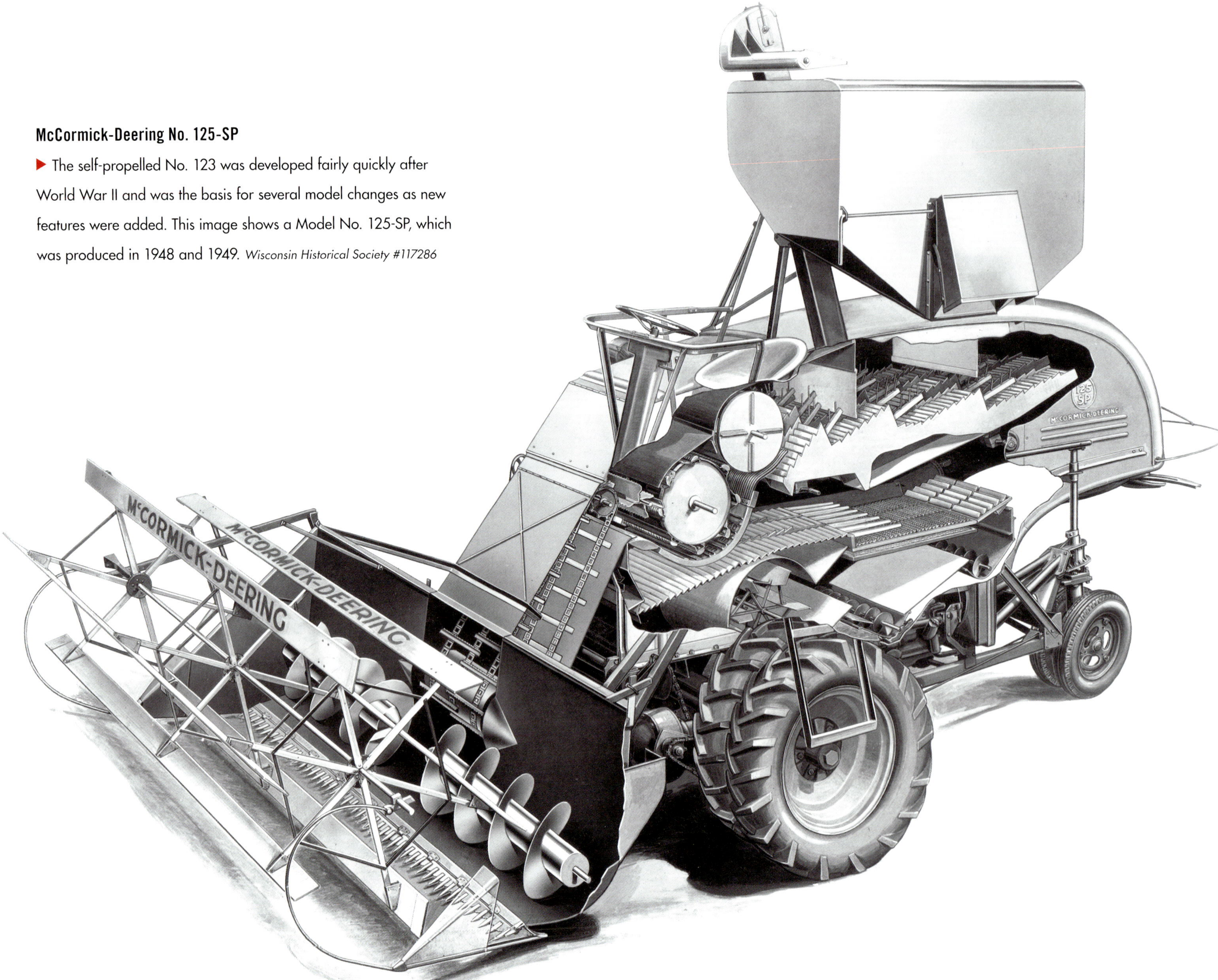

McCormick-IH 125-SPV Rice Field Special

◀ Model 125-SPV was built in 1950 and 1951, with the 125-SPVC produced in 1951 and 1952.

Wisconsin Historical Society #115115

McCormick-IH 1954 No. 127-SP

▲ A later iteration of the 123 was the No. 127-SP, produced from 1952 to 1954.

Wisconsin Historical Society #114892

McCormick-IH No. 127-SP

▲ More than 9,000 127-SPs were produced. *Wisconsin Historical Society #114930*

McCormick-IH No. 141

▶ The 141 replaced the 127-SP and represented a significant upgrade, with improved cooling and operator controls and comfort. The 141 was built from 1954 to 1957 and was available with 10-, 12-, and 14-foot platforms. *Lee Klancher*

McCormick-IH No. 141

▶ A corn head was released for the 141 in 1956. It didn't work all that well, but the innovation marked the beginning of an era in which corn harvesting would move to combines rather than tractor-mounted pickers. *Lee Klancher*

McCormick-IH No. 141

▼ More than 11,000 No. 141 combines were built. This survivor is operated by its owner, Daniel J. Tordai. *Lee Klancher*

McCormick-IH No. 141

▶ This image was taken during a 1955 tour of International Harvester's East Moline Works in East Moline, Illinois. This No. 141 combine engine is being tested and inspected. It was one of the last tests on the assembly line. *Wisconsin Historical Society #24778*

McCormick-IH No. 141 Hillside

◀ The 141 Hillside was the first machine with a system to level the platform both fore and aft as well as side-to-side. It was leveled hydraulically.

Dave Gustafson/Case IH

McCormick-IH No. 101

◀ The No. 101 was the first International Harvester combine designed with corn harvesting in mind. A No. 21 two-row corn head or a 10-, 12-, and 14-foot grain platform was available for the model. *Lee Klancher*

McCormick-IH No. 101

◀ The 101 was manufactured from 1956 to 1961, with more than 14,000 units built. This 101 is owned by Howard Ulrich, who is shown operating the machine. *Lee Klancher*

McCormick-IH No. 101

▲ The cockpit for the 101 was spartan, featuring temperature and oil-pressure gauges. *Lee Klancher*

McCormick-IH No. 101

▲ The 101 was powered by a 58-hp engine that drove a 27.25-inch cylinder. *Lee Klancher*

1960 McCormick-IH No. 151

▼ The 151 was a companion model to the 101 and was introduced in the United States in 1957. This No. 151 was manufactured in the U.S. and exported to France. *Jean Cointe Collection*

McCormick-IH No. 151 HS SP

▲ The 151 was also available in a hillside model, shown here. *Wisconsin Historical Society, #115080*

McCormick-IH No. 91

▶ No. 91 was built around a 41.5-inch cylinder and a unique steering system that allowed quick, tight turns. *University of Guelph*

THE SMALL SELF-PROPELLED COMBINES

International Harvester created a line of small combines that were built at its plant in Hamilton, Ontario. The machines were produced and marketed as small, affordable models ideal for the small or budget-minded farmer. The machines were built with a unique steering system.

The No. 91 was introduced in 1959 and produced until 1962, while the No. 93 replaced it in 1962 and was built until 1968.

The machines were quite popular, and the 93 evolved to become the No. 105. All of these machines are considered Class 1 combines. The small Class 1 and Class 2 machines were the best selling combines built by IH in the 1960s. That would change as farmers trended towards larger operations.

McCormick-IH No. 91

◀ The No. 91 featured an 8.5-foot platform and could be equipped with a No. 25 two-row corn head. More than 6,000 No. 91s were built between 1959 and 1962. Owned by Daniel J. Tordai. *Lee Klancher*

McCormick-IH No. 91

▲ The 91 had no steering wheel. The driver turned the machine with the two long levers that drove the left and right wheels, much like a crawler's drive system. International called the system "Magic Circle Planetary Steering." *Lee Klancher*

McCormick-IH No. 91

▲ The No. 91 was part of a line of small combines made by IHC, and it was built at the plant in Hamilton, Ontario. *Lee Klancher*

McCormick-IH No. 93 SP

◀ The No. 91 was replaced by the No. 93 in 1962. The No. 93 featured a three-speed transmission, a variable V-belt drive, and a standard automotive-style steering system in place of the planetary system on the 91. This example is owned by Max Armstrong. *Lee Klancher*

THE PULL-TYPE COMBINES

International Harvester kept a full-size pull-type combine in the line for decades but didn't do much to upgrade the technology. The first big machine was the No. 122, which was based on the No. 123-SP.

The machines were extremely popular around World War II, and the 122 was built from 1946–1949, during the heart of this boom. The hillside models were based on the 122, just with adaptations for uneven ground.

The subsequent models—the No. 64 through No. 76—were evolutions of the 122 design but not significant departures. By the time the No. 76 was on the market, demand had declined both due to fewer people using pull-behinds and the fact that the technology on the model lagged behind the competition.

Into the 1950s, pull-behind combines were very popular, but that would decline in the 1960s. According to IH engineer Don Murray's memoir, IH produced 8,050 pull-type and 3,117 self-propelled combines in 1955. By 1966, the company produced only 1,050 pull-type machines and 8,596 self-propelled units.

The No. 80 was a new machine that featured an auger-type platform, and was based on the No. 91 self-propelled combine. That was introduced in 1959, replaced by the 82 in 1966. The 82 was designed by IH engineer James Howell. The combine was a favorite with Midwest farmers, but it was discontinued in 1974.

1948 McCormick-IH No. 122

▲ The pull-type version of the 123 was this No. 122 combine, shown harvesting wheat behind a Farmall M tractor in 1948. *Wisconsin Historical Society #23922*

McCormick-IH No. 64

▲ The No. 64 combine was a pull-type unit that was either PTO-driven or configured with an auxiliary engine. This 1954 image shows a farmer harvesting grain with a McCormick Super WD-6 tractor and a No. 64 combine. *Wisconsin Historical Society #25232*

McCormick-IH No. 64

▲ The No. 64 had a 64.25-inch cylinder, which was very wide for the time. It could be equipped with a grain tank or a bagging device. *Wisconsin Historical Society #11488*

McCormick-IH No. 160 HS Pull-Type

▲ The big No. 160 Hillside combine replaced the No. 51 Hillside. The 160 featured a large 16-foot platform and was manufactured from 1951 to 1953. Only 388 were produced. *Wisconsin Historical Society #117331*

McCormick-IH No. 140 Pull-Type

◀ The No. 140 was offered with 9- and 12-foot platforms and could be configured with PTO drive or an auxiliary engine. A total of 5,626 units were built between 1954 and 1962. *Wisconsin Historical Society #115078*

McCormick-IH No. 150 Pull-Type

The No. 150 was a large-capacity pull-type machine that could be equipped with a 9- or 12-foot platform. This one is a Windrow Special. *Ken Updike Collection*

McCormick-IH No. 76

The No. 76 replaced the No. 64 pull-type and was produced from 1955 to 1958. More than 20,000 units were built. *Lee Klancher*

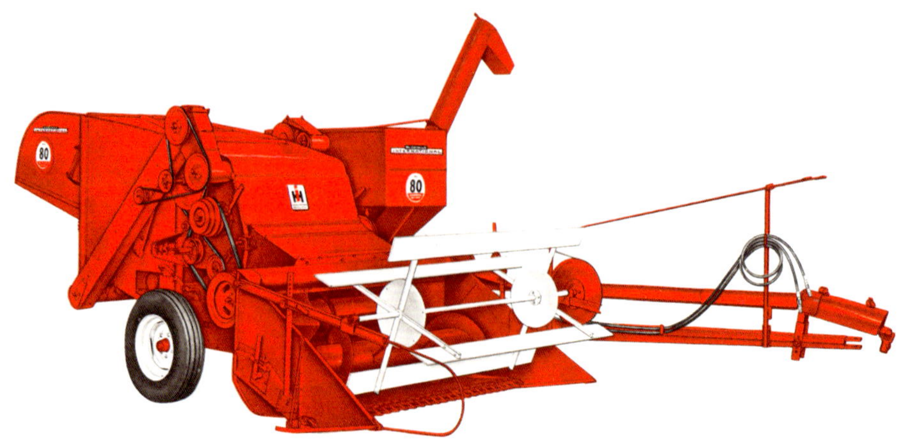

McCormick-IH No. 80 Pull-Type

No. 80 pull-type had a 7-foot platform and a 42-inch separator that fed a 26-bushel grain tank. The No. 80 was also available as an Edible Bean Special. *Ken Updike Collection*

International No. 82 Pull-Type

▲ The No. 82 replaced the No. 80. More than 3,700 units were produced from 1966 to 1974. "The delightful little 82 had been the favorite of Midwest farmers till they began to consider corn harvesting with the combine," Don Murray wrote of the model. This one is owned by Max Armstrong. *Lee Klancher*

J. I. CASE COMPANY 1942–1961

The Case side of the Case IH legacy continued through the middle of the twentieth century. Case continued to build combines from 1942 to 1961, as well as its line of tractors. During World War II, the company built military equipment for the U.S. government.

The time following the Second World War was the best time in history to be selling tractors, and the company grew its line of equipment through this era. While it remained one of the smaller players in the industry, anyone in the game at that time was able to do well.

In 1953, J. I. Case produced the last threshing machine, as it was replaced by the combine.

Case Combine

▼ Case continued to manufacture combines in the 1940s and 1950s. *Case IH*

1954 Case Model 120 Rice Special

◀ Model 120 self-propelled combines were built between 1954 and 1958 and featured power steering. This Rice Special version had larger floatation tires, a planetary-drive front axle, and dual disc brakes. *Case IH*

Chapter Three

1962–1977 THE BIG TOUGH RUGGED LINE

By Lee Klancher

"But even when everyone was pitching in, we never got out from under all the work. I loved that ranch, though sometimes it did seem that instead of us owning the place, the place owned us."
—Jeannette Walls, *Half Broke Horses*

Mary Conger and her husband's large Kansas dairy farm was one of the largest-grossing farms in the area in 1960 and used the latest technology on nearly every single front. Even so, the pair struggled to make ends meet.

"When the first slash in farm prices came six years ago, we doubled our milking herd in an effort to increase gross income so that, in turn, we could meet our fixed charges—such as interest and taxes—and the rising cost of things we must buy," Conger wrote in the *Saturday Evening Post*.

"We built a new labor-saving milking parlor.... But then came years when our crops were cut by drought, hail, and wet weather, and we fell behind on the feed bill for the cattle. In good years, we struggled to catch up. We tripled the milking herd. Milk prices declined further. Costs went on up. We were on a treadmill, always running faster just to keep the same place."

Mary Conger's widely read magazine article captured the rapid changes and difficulties facing the farmer due to mechanization and social change, and it was later read into the *Congressional Record*.

Farming is tough enough due to the inevitable price fluctuations and the fickle moods of Mother Nature. During our transformation from an agrarian society to an industrialized world, the changes required to survive as a farmer were massive. Yields had to rise, and so did the acres worked. Fewer and fewer people were feeding the world, and technology played a key role in doing that.

Simply embracing technology and growth wasn't enough. Farmers needed to be shrewd to survive these times.

Knoxville farmer John West was one of those. He worked rented land and earned enough money to purchase 126 acres of land in Blount County, Tennessee, in July 1937. John was able to harvest 33 bushels of wheat per acre, a respectable figure at the time.

Over the years, he invested thoughtfully, relying heavily on borrowed or hired equipment whenever possible and

Modern and Efficient

◀ During the 1960s, International made upgrades to improve the customer experience. Dealership designs were modernized and a new network of parts depots was built that allowed 24-hour delivery of any service part. Historically strong from a sales perspective, Harvester continued that trend with an increased focus on training its sales force to understand the plights facing the farmer. In this image, Jim Brosnahan provides instruction at the company farm at Tifton, Georgia, in June 1962. *Wisconsin Historical Society #91613*

Technology on the Farm

◄ By 1960, America had transitioned from an agrarian nation to an urban one, with only 8.3 percent of the population working on the farm. One farmer's work fed an estimated 25.8 people, and most farm work was done with powered machinery. This farmer is looking at data about his dairy cows' output on January 27, 1961, in Madison, Wisconsin. *Wisconsin Historical Society #6996*

Vietnam War

▲ The early 1960s were a boom era in the United States, but things would change with the escalation of the Vietnam War in 1965. The conflict proved a drain on the economy. America was no longer primarily agrarian, but this wasn't the case in much of the rest of the world. Places touched by war, in particular, were set back decades by destroyed cropland and decimated workforces. *Wisconsin Historical Society #55004*

diversifying by raising corn for feed and cash as well as wheat, poultry, and hogs. He also trapped rabbits and muskrats in late fall and winter, selling furs to earn a few extra dollars.

He borrowed money only on occasion and generally to add cattle or land.

"After the first year, I started expanding. I bought cattle at Knoxville in the sale, 2 cents a pound. Borrowed the money at the Bank of Maryville to pay for them. I'd go up there and talk to them [bank presidents] just like I'm talking to you: tell them what my ideas was and that we needed some money, and I was never turned down."

One banker told West, "All I know to do is let you have it. Everything you ever tried works."

For a farmer like West, a new paint job and high-tech advertisement would not be enough to convince him to part with his hard-earned money. The line-up International Harvester offered in the 1960s and 1970s was efficient and pretty enough to get some to part with their money, but it wasn't world-class.

Rich McMillen was born in Dewey, Oklahoma, in 1941 and grew up on a family livestock and crop farm in southeast Kansas. He started operating machinery when he was 12 years old and went on to become an engineer working on red combines from February 1966 until 2000. He worked as a design engineer in the late 1960s, and the farm boy from Kansas understood that the machines the company had to offer weren't efficient enough to get men like John West to invest in them.

"With our 815 and 915 combines, we had some problems," McMillen said. "We probably were down to 10 to 15 percent market share because there was other things in the company like, I believe, the problems we were having with the tractors in those days, the 560 and 660 and some of the differential problems that we had with them, hurt the whole company and our combine market.

"We didn't have anything special with the 815 and 915 combine. We basically had what everybody else had. I believe that knowing what we could do in rice was a stroke of luck. We didn't develop it in the beginning for the rice field. We were developing it primarily for corn, wheat, and the soybean harvest. What we found out after we released it for production, is that we got the reputation of running good when the sun shines, but when the cloud comes over, you'd better pull it in the shed. That's the way it was in the rice and some of the tough wheat and soybeans."

Thankfully for International Harvester, McMillen's coworkers had been secretly developing a technology that was more efficient. That development started in the 1950s, and McMillen would join in to help out with the machine in 1972.

In the meantime, red combines played a supporting role as the modern world transformed the farm.

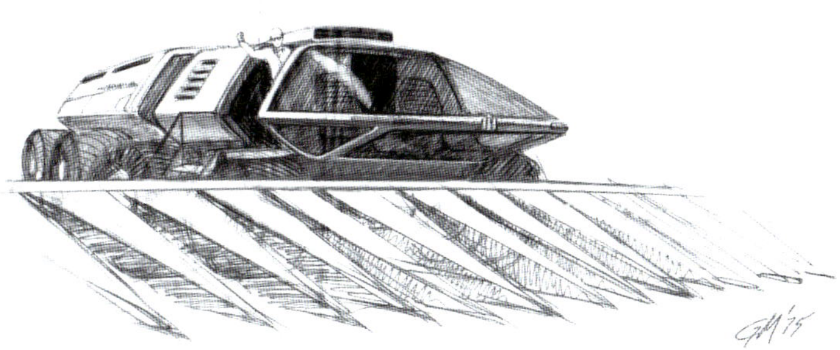

Solar Division

▲ One of Harvester's smart diversifications in 1960 was the purchase of the Solar Division, which built turbine engines and high-technology components, including the beryllium container shown above sitting on the lunar dust during the Apollo 12 mission. *Wisconsin Historical Society #95228*

03 SERIES

The Model 101 was a successful new machine for IH, and its emphasis on corn harvest gave the company a nice lift in the market. As a result, development continued on an entire line of new machines that were released for 1962 production.

The 303, 403, and 503 combines had a host of new features designed to make them more comfortable, easy to use, and to improve visibility for the operator. They also featured more horsepower and capacity, as well as a lower height to make them easier to transport.

The larger 403 and 503 combines featured hydrostatic drive, which provided variable speed control and eliminated the fussy variable V-belt used on older models. The hydrostatic drive was designed by IH engineer Gene Krukow, who owned a farm in north central Iowa and was trained in hydraulics during a stint with the U.S. Air Force.

According to IH engineer Don Murray, the 303, 403, and 503 accounted for 81 percent of IH combine sales in 1966. The company produced 11,167 combines in 1966, which brought in $87 million in revenue. The revenue was good, but other brands were doing better as the 03 Series was aging.

"But by 1966," he wrote, "they were getting a little long in the tooth and the competition was giving us trouble." He went on to say that IH's market share was at 15 percent at that time, down from 22 percent in 1962.

"Of course management was getting anxious," Murray wrote, "and good old Don wandered into this situation."

The situation he walked into? International Harvester desperately needed a new combine in the late 1960s. Thanks to a stubborn Swedish-born IH engineer with a penchant for ignoring management directives, Don would find he had an ace up his sleeve.

403 Prototype Drawing

▶ The 403 was the replacement for the 181 combine. This drawing of the proposed 403 is dated June 16, 1959. *Wisconsin Historical Society #115079*

1. "Flip-up" type upholstered seat – fully adjustable.
2. Operator's platform raised.
3. Farmall tractor type steering wheel positioned for operator comfort.
4. All controls redesigned for improved operator comfort.
5. Low grain tank permits viewing from seated position.
6. 65 Bushel grain tank.
7. 90 Horsepower engine.
8. Positive clutch in engine separator drive.
9. Rear access ladder to engine placed on left hand side for operator's convenience.
10. Lowered overall height for ease of truck transport.
11. Separator lengthened 14" and hood sides made deeper.
12. Main sill of strong angle construction eliminates need for unwieldy truss rod construction.

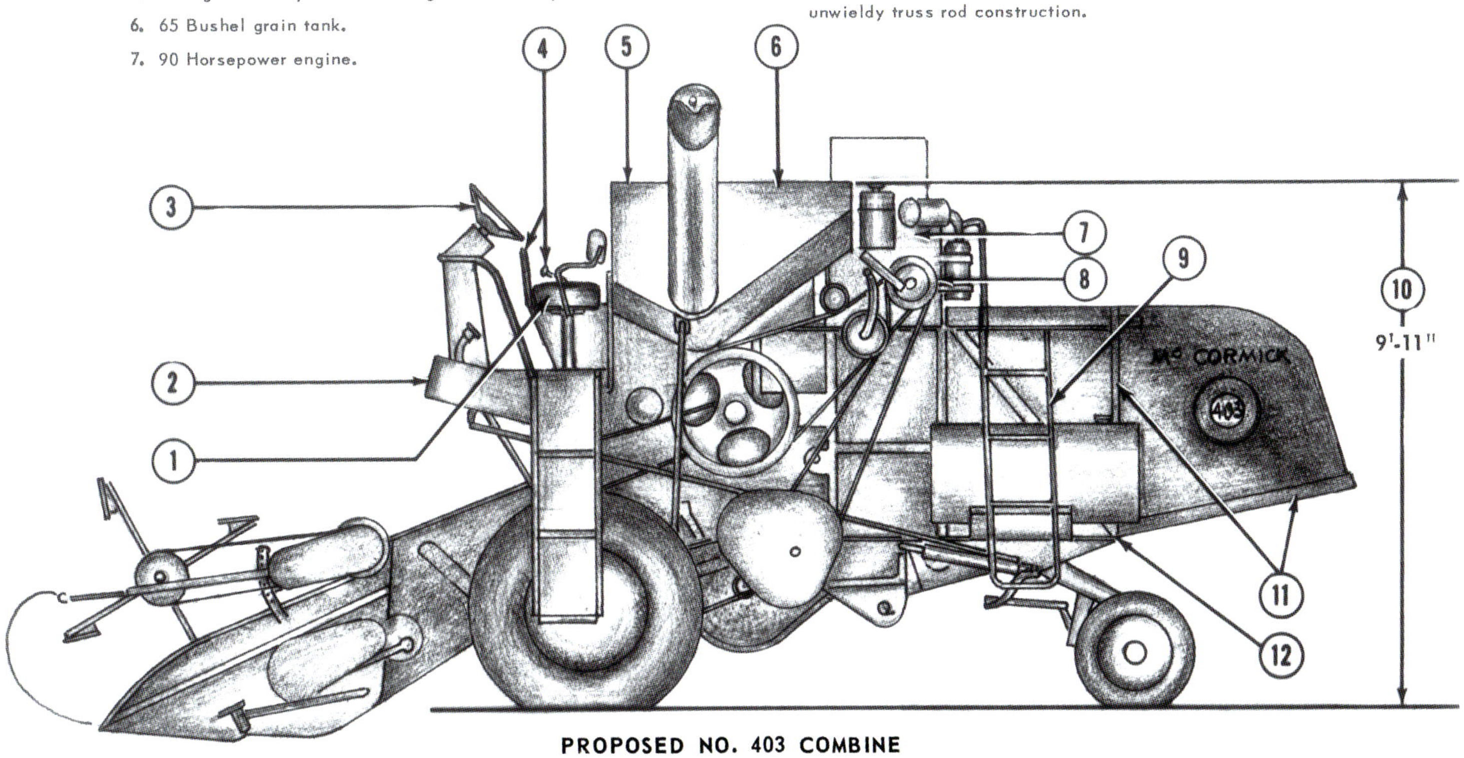

PROPOSED NO. 403 COMBINE
(Typical – other machines similar in side views.)

Specifications for New Line

◀ This drawing shows the new features that IH intended to introduce on its new-for-1962 line of combines. *Wisconsin Historical Society #117400*

McCormick-International 303

◀ Introduced in 1962, the No. 303 replaced the No. 101 combine. The 303, like the 403 and 503, was equipped with a 22-inch-diameter cylinder. The 303's cylinder was 30 inches long. Owned by Daniel J. Tordai. *Lee Klancher*

McCormick-International 403

▶ The 403 was the middle of the new lineup, replacing the No. 151 combine. The line featured much improved operator controls and comfort, and a Farmall-type steering wheel. Hydraulic power steering was standard on the 403 and 503. The grain tanks were lowered and separated to the right and left, which lowered the center of gravity. This 403 is owned by Daniel J. Tordai.

Lee Klancher

McCormick-International 503

▶ The No. 503 replaced the No. 181 combine in 1962. The 503 was powered by a 106-hp engine and had a 106-bushel grain bin. The big 503 could be fit with a 20-foot grain head—IH's largest. It also could handle a six-row corn head.

Wisconsin Historical Society #115103

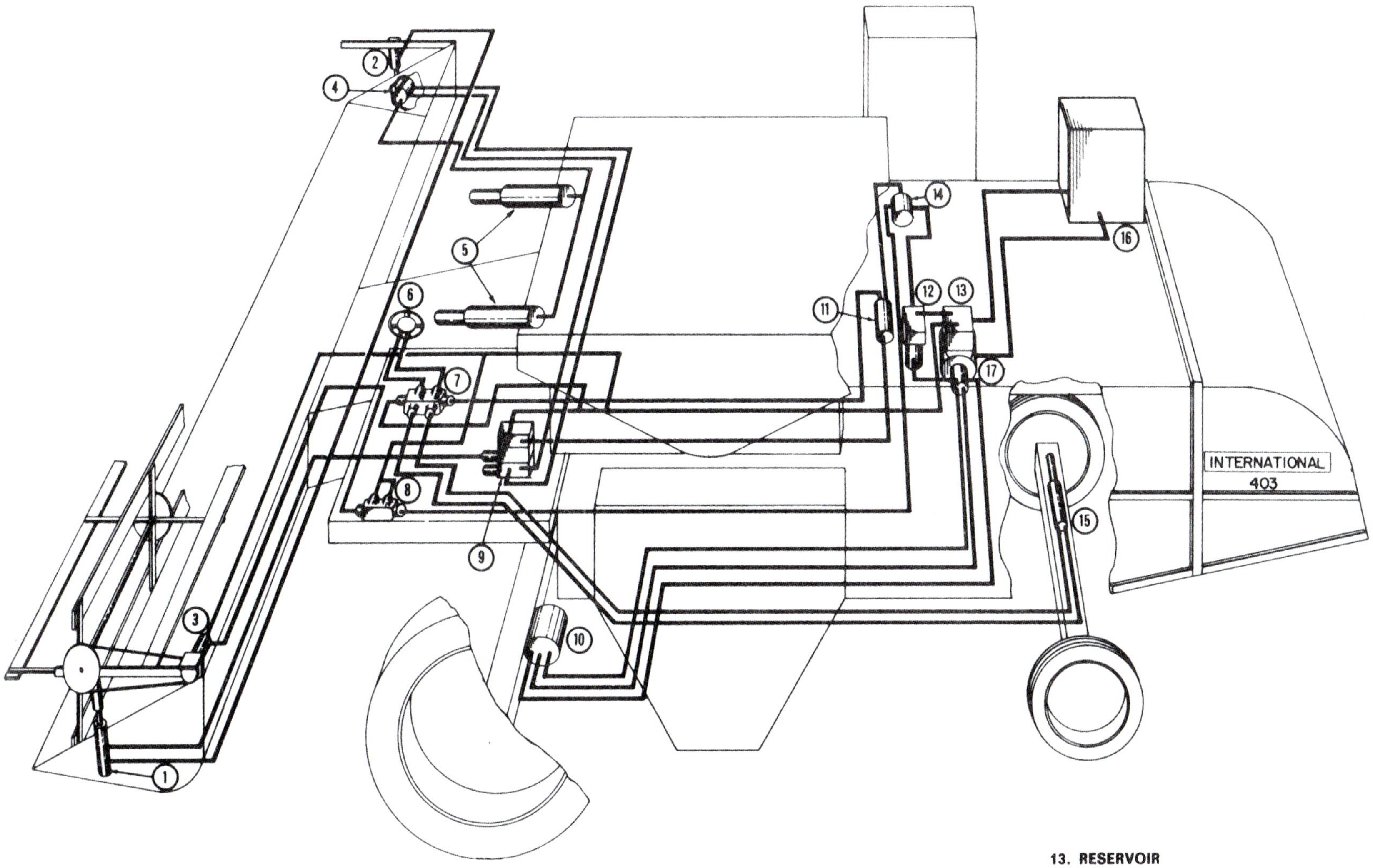

1. REEL LIFT MASTER CYLINDER
2. REEL LIFT SLAVE CYLINDER
3. REEL DRIVE MOTOR
4. AUTOMATIC PLATFORM FLOW CONTROL VALVE
5. PLATFORM LIFT CYLINDERS
6. STEERING HAND PUMP
7. STEERING CONTROL VALVE
8. REEL DRIVE SPEED CONTROL
9. REEL & PLATFORM LIFT VALVES
10. HYDROSTATIC GROUND DRIVE MOTOR
11. STEERING FLOW DIVIDER
12. GROUND DRIVE OIL FILTER
13. RESERVOIR
14. DOUBLE PUMP
15. STEERING CYLINDER
16. OIL COOLER
17. HYDROSTATIC GROUND DRIVE PUMP

403 Hydraulic System.

16 Looking at the illustration above we can see that the hydraulic system is not centralized but spread out over a major portion of the grain harvesting machine. The steel tubing used to interconnect the hydraulic components is an effective heat exchanger. Systems operating at 15 to 23 G.P.M. oil flow can use reservoirs as small as three gallons. Additional hydraulic applications have been tested and will soon be available on International grain harvesting machines.

New Hydraulic Systems

▲ The 403 and 503 were equipped with sophisticated hydraulic systems in the mid-1960s. Automatic platform-height control was introduced in 1965. The system also introduced hydraulic reel lifts and draper pickup drives. The steering was all-hydraulic and dubbed "hydrostatic" steering. The drive to the wheels was also hydrostatic. The 403 and 503 had all of these available as standard or optional equipment; a few of the hydraulic features were available as options on the other machines in the line as well.

Wisconsin Historical Society #117409

McCormick-International 403 Hillside

▼ The No. 403 Hillside was eventually available as a fully self-leveling machine. This meant the platform could be leveled fore and aft as well as side to side. The system was fully hydraulic and featured a number of innovations unique to the industry.

Wisconsin Historical Society #115101

McCormick-International 403 Hillside

▲ If the hydraulic system failed or a hose ruptured, the hydraulic valves on the fully self-leveling 403 would lock in place. This safety feature prevented a hydraulic failure from causing the machine to topple if the failure occurred while on an extreme slope. *Wisconsin Historical Society #115105*

McCormick-International 403 Hillside

◀ The No. 403 was the hillside model in the series. The machine was available with a side-leveling system as well as a hydraulic fully self-leveling system.

Wisconsin Historical Society #115104

McCormick-International 203

▲ The 203 featured a 41.25-inch-wide cylinder, like the 93. Unlike the 93, it was equipped with twin grain tanks, dubbed "saddle tanks," that lowered the center of gravity. During five years of production, 5,135 203s were built. This machine is owned by Daniel J. Tordai. *Lee Klancher*

McCormick

McCormick-International 105

◀ The No. 93 was replaced with the 105 as the International Class 1 combine. This example is owned by Daniel J. Tordai. The farm in the background is owned by Tim Wilhemi, as is the Case IH 1660 combine parked next to the barn. *Lee Klancher*

McCormick-International 105

▼ The 105 featured an L-shaped grain tank and an improved straw-rack system.

Lee Klancher

McCormick-International 105

▲ The 105's controls showed significant technological improvement over the 101. Note the platform and reel remote controls. *Lee Klancher*

International 105

▶ The 105 was available in two different paint schemes. Later models had a white top, as seen on this one owned by Daniel J. Tordai. *Lee Klancher*

International 205

▶ The 203 was upgraded to the 205 and was produced from 1967 to 1971. This one is owned by Debbie Tordai. *Lee Klancher*

International 205

◀ The 205 was the last of the International Class 1 combines. The small class was very popular in the middle and late 1960s, taking 30 percent of the market in 1966. That dropped off starting in 1967, and by 1971 only a few thousand Class 1 combines were sold in the United States. International discontinued the line when the 205 was taken out of production in 1971. *Lee Klancher*

402 Pull-Type

◀ The 402 Pull-Type offered all the features of the 403 and provided a higher-capacity pull-type than IH had built in the past. The model was introduced in 1966 and produced until 1971.

Wisconsin Historical Society #117289

15 SERIES

The 15 Series started out about as poorly as possible. The opening shot was the 315, which was based on the old 203 / 205 chassis. The machine was designed by IH engineer Charley Hyman, who was in his fifties during the project. He and his brother, Benjamin, had been the lead engineers for corn pickers for many years.

"For some reason I became suspicious of the overall design of the 315 very early and proceeded to give Charley a hard time, but maybe not hard enough," IH engineer Don Murray acknowledged. "The 315 was quite an experience."

International Experimental Combine

▲ This intriguing image shows what appears to be a 315 combine badged as a 415. The 415 was not produced, so this machine is most likely an early production version of the 315.

Wisconsin Historical Society #115102

Introduced in 1967, the machine was unloved, sold poorly, and was considered a company embarrassment.

Former IH executive Ralph Baumheckel wrote that senior executive (and eventual CEO) Brooks McCormick personally stopped in to ask the chief engineer for grain harvesting what the problem was with the 315. The engineer explained that the machine didn't work well in corn, and suggested they build a replacement as soon as possible.

Brooks was a smart, motivated leader, and he apparently listened. The ill-fated project was scrapped and replaced with a new machine, the 615, which was introduced in 1971.

Thankfully for International Harvester, the 815 and 915 weren't such disastrous launches. Scheduled to launch in 1968 and initially designed with twin grain tanks, the 815/915 machines were delayed when pretty much everyone from the product committee to the engineers realized that the new machines needed to have a single grain tank mounted high on the rear of the machine.

The change delayed matters, but the results proved better than the horrid experience with the 315. The 715 was the last machine added to the line, and it provided an intermediate step in performance and budget between the small 615 and the big 815 and 915.

The 15 Series introduced a host of new features and was able to keep IH's market share in the combine at a reasonable level. The line, however, had a reputation for working well in dry conditions and horribly after a drop of rain.

This series was not a game-changer. More would be needed for IH's combine line to stand out in the crowd.

International 315

▶ The 315 was the first of the new 15 Series combines and was actually an improved version of the 203. The 315 was introduced in 1967 and featured a 20-inch-diameter cylinder, 70-bushel grain tank, and a 72-hp engine. This 315 is owned by Jerome Jecha. *Lee Klancher*

INTERNATIONAL
315

International 815 High-Profile

▲ This 1969 advertising poster shows the brand-new 815. This version of the 815 (and also the 915) was produced until 1974, when a low-profile variant was introduced that was shorter and designed to fit through a 12-foot-high shed door. The older models can be distinguished by the cab that sticks high above the grain bin and by the sheet metal under the front of the cab that is painted white and emblazoned with a black "815" sticker. *Wisconsin Historical Society #51379*

International 815 High-Profile

▲ The all-new 815 brought a number of new features to the line, including an air intake that was mounted high on the machine. On the older models air was drawn from underneath the rear of the machine (along with a lot of dust and debris in the process). The new location provided cleaner air. *Gregg Montgomery Collection*

International 815 High-Profile

▲ The protruding cab is a distinctive feature of the early 815 high-profile models, and it's clearly visible in this image. Note how the cab sticks up well above the grain tank deck. *Gregg Montgomery Collection*

International 615

◀ A much-improved version of the venerable 303, the 615 was introduced in 1971 as a replacement for the unpopular 315. The 615 was available in belt or hydrostatic drive and featured new quick-attach corn and grain heads. Available engines were an IHC C-263 gas and a D-282 diesel. This one is owned by Daniel J. Tordai.

Lee Klancher

International 615

▶ The 615 was equipped with an 83-bushel grain bin and could handle 10-, 13-, or 15-foot grain heads and two- or three-row corn heads. Owned by Daniel J. Tordai.

Lee Klancher

International 715

▶ The 715 was introduced in 1971 and was designed to replace the 403. The 715 had a large 93-bushel grain bin and could handle the same heads as the 615 plus the 20-foot grain head and four-row corn head. The 715 was produced until the 1420 came out, and the value-priced machine was a popular choice until the end. This is owned by Daniel F. Tordai.

Lee Klancher

International 715

▼ The 715 came with either the IH C-301 gasoline engine rated for 107 hp or the IH D-301 diesel rated for 95 hp. Owned by Daniel F. Tordai.

Lee Klancher

914 Pull-Type

▲ The 914 was the largest pull-type combine on the market when it was introduced, and it featured a 150-bushel grain tank. The 12-inch unloading tube could crank out 114 bushels per minute. *Ken Updike Collection*

453 Hillside

◀ The 453 was a four-way leveling machine that could be ordered with 16.5- or 18.5-foot grain headers.

Ken Updike Collection

453 Hillside

◀ The 453 was the only four-way leveling machine on the market. *Ken Updike Collection*

International 815 Low-Profile

▶ The 815 and 915 were loved for their features but hated for their height, which made them difficult to fit in the shed. International upgraded the 815 and 915 dramatically, introducing the low-profile models in 1974. This image shows a farmer harvesting corn with an International 815 Low-Profile Hydrostatic combine, an International 574 tractor, and a farm wagon. The 815 could be equipped with a V-8 gas engine or a six-cylinder diesel. *Wisconsin Historical Society #8536*

International 815 Low-Profile

▶ The two distinctive features of the late-model low-profile combines are clear in this photo: the cab is only slightly higher than the grain tank, and the sheet metal under the cab is red with white number lettering. *Gregg Montgomery Collection*

International 915 Low-Profile

▶ The new-for-1974 low-profile models featured double catwalks, an improved cylinder, a quick-attach feeder housing to more easily swap heads, and special editions for grain and corn, as well as rice and edible beans. Owned by Tim and Christine Wilhelmi. *Lee Klancher*

INTERNATIONAL
HYDROSTATIC
915

Museum of Science and Industry 1975

▶ Visitors examining a 915 hydrostatic combine in the International Harvester exhibit at the Museum of Science and Industry in Chicago, June 1975. *Wisconsin Historical Society #41386*

International 915 Low-Profile

◀ The 915 proved to be a worthy machine once the second generation machine was created, but was released in an era when too many improvements were done late in the game. *Lee Klancher*

INTERNATIONAL HARVESTER
OWNERS REGISTRATION

HOMME BROTHERS
RR 1
THOMPSON ND 58278
338841551 EXP DATE
915WDRW SPEC 12 21 74

157 0118 U003372 1165516

Mint-Condition 915 Windrow Special

▲ This 915 was discovered in 2014. It's a late-model version, as evidenced by the IH registration card, with very few hours on the clock. It is owned by Darius Harms, the organizer of the Half Century of Progress Show held in Rantoul, Illinois. *Jill's Creative Expressions*

J. I. CASE COMPANY 1962–1972

In the 1960s, the J. I. Case combine line was able to hold 10-15 percent of the market, according to Don Murray's memoirs. This was a respectable figure at the time, and the combine line seemed to be moving in the right direction for the company.

"Case [combines] had a sort of conventional layout," Murray wrote, "tank on top, engine in rear up top, but somehow they always looked unfinished to me, just an opinion."

J. I. Case never built a Class 5 machine, and the company stopped combine production in 1972.

Case Model 600

▶ The Model 600 was introduced in 1961 and produced until 1965. Options included gasoline or diesel engines. Hydrostatic power steering was standard. *Case IH*

1959 Case Model 1000

▶ The 1000 featured a 42-inch cylinder and four straw walkers. A 10-foot header and Case 718 gasoline engine were standard; options included diesel and LP engines as well as a 14-foot header, corn head, bagging platform, and an electrically operated header clutch. *Case IH*

1959 Case Model 660

▲ The 660 was powered by a Case 201-ci gas engine. Options included a Case diesel engine, an Edible Bean Special, and larger headers. Production was suspended in 1972 and resumed after the Case IH merger in 1985. *Case IH*

1959 Case Model 960

▲ The Case 960 was powered by a Chrysler 225-ci six-cylinder engine and built from 1966 to 1972. *Case IH*

Old Abe Retires

▲ In 1969, Case changed the company logo—Old Abe no longer carried the flag for the company. *Case IH*

Chapter Four

1954–1977 DEVELOPMENT OF THE AXIAL-FLOW COMBINE

By Lee Klancher

"Some organizations are very strong on early analysis and objectives but, realizing the difficulty of the endeavor, never really tackle the job. East Moline Engineering was the opposite, they were not real sure where they were going but worked like hell to get there. . . . Along in the middle they figured out where 'where' was and succeeded."
—Don Murray, Memoirs about creation of the Axial-Flow combine

Elof Karlsson was born and educated in Sweden. He had a heavy accent, a brilliant mind, and a propensity to do things his own way, even as a young man.

His job interview with Harvester took place in Chicago at Tractor Works and became the subject of company lore. As the story goes, when the interviewer walked into the room, he found Karlsson poring over a cutaway view of a transmission and differential in a brochure.

"Why the hell would you design it like that?" were the first words out of Karlsson's mouth.

The interviewer was shocked and shot back, "Well, how would you do it?"

Karlsson then proceeded to sketch out the correct way to design the transmission. The correct way—always—being the Karlsson way.

The design was good. Karlsson was hired on the spot.

Karlsson became a creative force at Harvester. His name appears on numerous patents, with much of his focus on harvesting equipment.

His ability to do complex math in his head was legendary.

When calculating loads and strains, Karlsson would scribble down the formula, work out the math, and have the answer in a few minutes. The fact that he had those formulas in his head was a source of amazement for his fellow engineers.

He was intensely focused on crafting new machines and interested in progressive technology. He had little time, however, for directives that didn't suit his needs.

Ken Johnson started as a young engineer at Harvester in 1956. Elof, by that time, was in his mid-fifties. He was testing a variety of wild ideas and strange designs. He soon discovered that Johnson was a quick study and a useful assistant. He put

Elof Karlsson

◀ Elof Karlsson was a brilliant, outspoken International Harvester engineer whose vision was the driving force behind the earliest development of the Axial-Flow combine. He is shown here in the 1950s, in front of his new 1953 Cadillac.

Elof Karlsson Family Collection

him to work on various projects, including calculating how much air was required to lift a single kernel of corn.

One day Elof felt he needed a change of venue. He didn't drive, so he called Johnson for help.

"Do you have a pickup down there?" he asked.

"Yeah," Ken said.

"Come pick me up. I've got to talk to you. We're going to go downtown and have a cup of coffee."

Johnson drove them to a café, and Elof proceeded to sketch away madly and make lists of things for Johnson to acquire and do.

Eugene Mel Van Buskirk

▲ Harvester design engineer Eugene Mel Van Buskirk played a key role in developing the Axial-Flow combine. He was born on July 28, 1912, and started work at Harvester in 1934. A gifted draftsman, he created many of the late-1970s engineering sketches for the Axial-Flow's development.

Mel Van Buskirk Family Collection

A few minutes later, Chief Product Engineer Ray Barkstrum walked into the café. Engineers were not allowed to leave the plant at that time, and the product engineer had witnessed Elof's escape.

He wasn't pleased.

Elof calmly kept working. He eventually told Barkstrum what he and Johnson were working on. Then he turned to Johnson.

"Let's see," Elof said. "Where were we before we were so rudely interrupted?"

Elof didn't play politics. All he cared about was his designs.

Ernest "Mel" Van Buskirk started working as an engineer at Harvester about the same time as Johnson. He spent some of his early days in the brand-new engineering facility at Hinsdale, Illinois. The center was built to centralize Harvester engineering and was a showcase building equipped with state-of-the-art test equipment and facilities.

It was also close enough to Chicago that top brass could keep a watchful eye on product development. In the 1950s, those watchful eyes produced some god-awful disasters, the most notable being the rush to production of the 560 tractor that resulted in a raft of recalls and an industry-wide black eye.

The combine engineers were still working in old facilities located in East Moline, not far from the East Moline Plant where combines were built. The location kept the engineering team far enough from the Chicago brass to stay off the radar.

Combine engineering was a relatively small group located in a giant plant. That meant they had to maintain proper relations with manufacturing to survive, sometimes at the expense of the relationship with Hinsdale, not to mention the brass in Chicago.

On occasion even today, the combine engineering team is known as "those rebels from East Moline."

Those rebels being off the radar is perhaps the only reason the Axial-Flow combine *exists*.

Van Buskirk was transferred from the modern digs in Hinsdale to the dingy, occasionally rebellious halls of East Moline. When he started at the East Moline Product Engineering Center, the environment was relaxed.

"I had come back to East Moline after I transferred to Hinsdale and we were sitting in the office down there one rainy day," Van Buskirk recalled. "The chief engineer hadn't given me a project to work. He just asked me to find something to do."

That suited Van Buskirk just fine. What he found was Elof Karlsson.

Karlsson was working on revolutionary, ambitious projects. Van Buskirk thought that was grand. The two hit it off and worked together on a variety of machines.

"We were both . . . dreamers. We could sit down and we'd try to out-dream each other," Van Buskirk said. "We made sketches and we both liked to doodle when we were talking. We would go ahead and try to do all the things that you might want to do."

The first major project Van Buskirk and Karlsson worked on together was a beast of a sugar cane harvester. The thought at the time was sugar cane should be cut a few inches under the soil, so the harvester dug into the ground below the plants. Harvesting the high-moisture crop and cutting into the soil required a lot of horsepower—two Chrysler V-8 engines powered the raucous machine.

Testing took place in Cuba, Puerto Rico, and Peru. As Van Buskirk recalled, the cane harvester was a flawed demon. "We could cut, clean, and load a ton of cane a minute, but we was leaving about 20 percent on the ground," he said.

The idea for rotors came about on Karlsson's and Van Buskirk's next project: a rotary corn sheller. The machine worked extraordinarily well, and the two visionaries quickly understood the implication: they could build a much faster corn harvester using rotary separation.

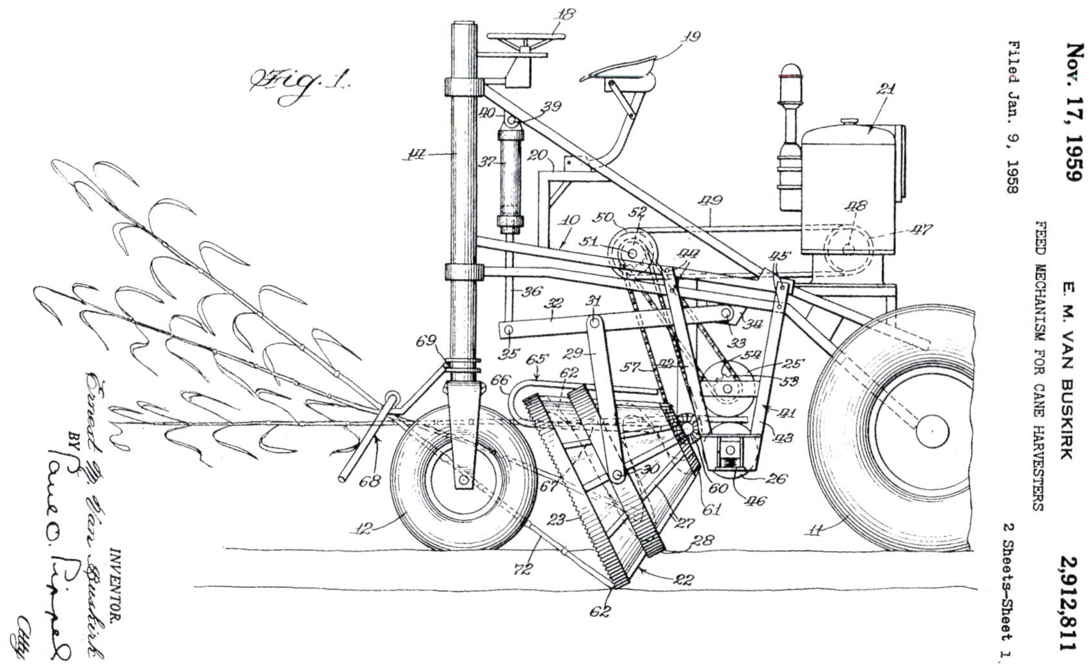

Cane Harvester Patent Drawing

▲ One of the first projects on which Karlsson and Van Buskirk collaborated was the cane harvester. Van Buskirk filed this patent for the machine's feed mechanism. *U.S. Patent Office*

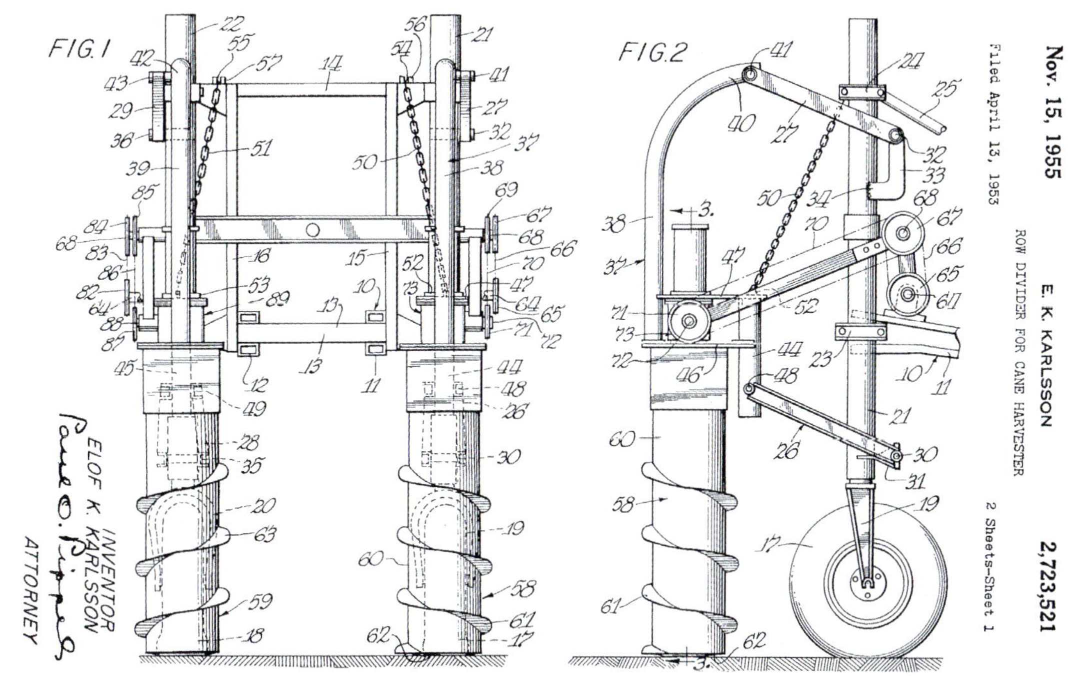

Cane Harvester Patent Drawing

▲ Karlsson's patent for the cane harvester provided a way to separate rows. *U.S. Patent Office*

EARLY ROTARY COMBINE DESIGNS

The fundamental advantages of rotary threshing were understood in the 1950s, at least by progressive innovators such as Karlsson. The concept of separating grain using centrifugal force had been around since 1772, when a design was created by Swedish innovator Sven Ljundqvist. The idea was to replicate the motion of a human hand flailing the grain free from the stalk; a wide variety of mechanisms were designed to do this.

American innovator Curtis Baldwin was one of the early proponents of a rotary combine, and he founded several companies during the 1930s in an effort to bring his rotary thresher to market.

Baldwin's efforts proved in vain. The concepts may have been understood, but making a rotary thresher that actually worked was something the brightest minds of the early twentieth century were unable to accomplish.

One of the companies that came closest was Harvestaire Inc. of Sacramento, California. Under the direction of H. D. Young, Harvestaire set out to produce a commercially viable rotary combine that used forced air in the process. By 1955, the company had three units being tested in the field. These experimentals showed promise with speedy harvesting in some conditions but were overly complicated, underpowered, and mechanically unsound.

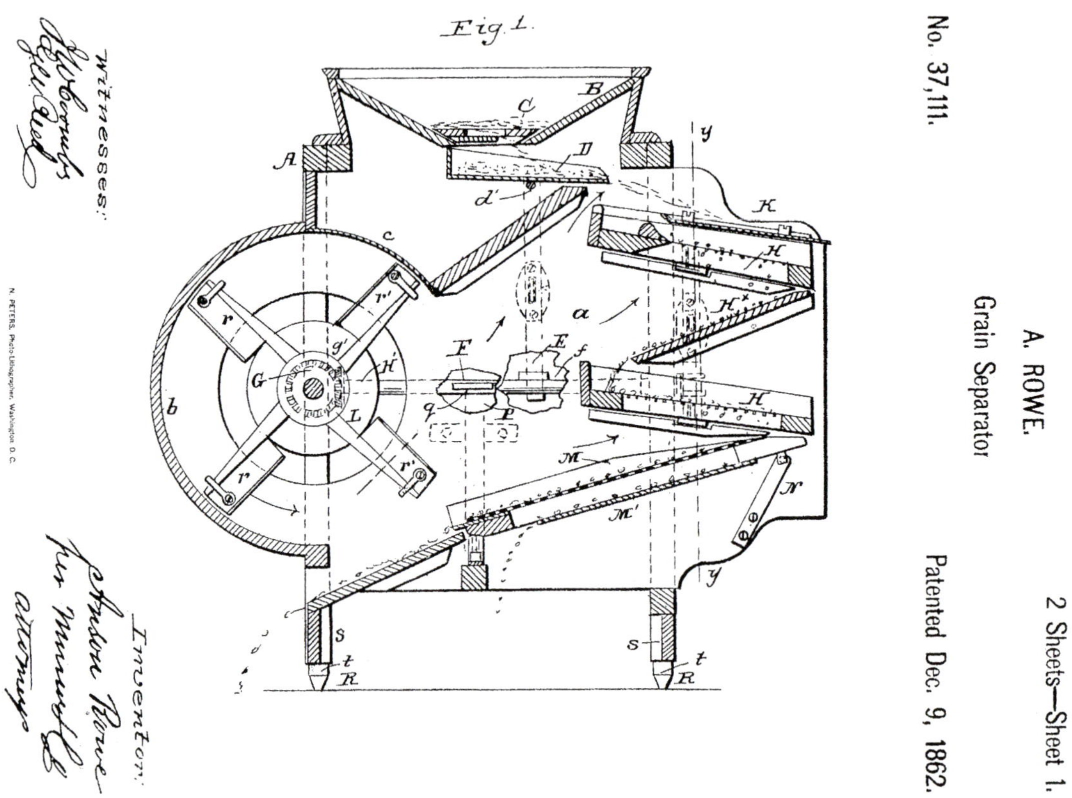

Grain Separator, Patented 1862

▲ The idea of separating grain using a rotating blade or cylinder is an old one, as evidenced by this patent. Anson Rowe of Atalissa, Iowa, provided a convoluted description of his idea for improved grain separation. The concept's rotating blades and circulating air are the foundation of rotary separation, and his patent was cited more than 100 years later in Karlsson's patent for a rotary separator. *U.S. Patent Office*

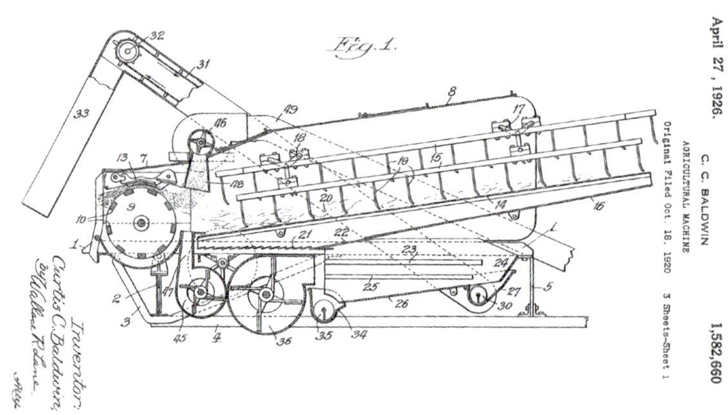

Curtis Baldwin

▲ One of the most tragic and interesting stories in early combine history is that of Curtis Baldwin, who developed a series of innovative machines and tried to market them with a series of ultimately doomed companies. Baldwin's designs were sold to Massey-Harris shortly before he died. Today, fewer than a half-dozen Curtis harvesters are in private collectors' hands. This patent is for one of his early combines that used rotary separation. *U.S. Patent Office*

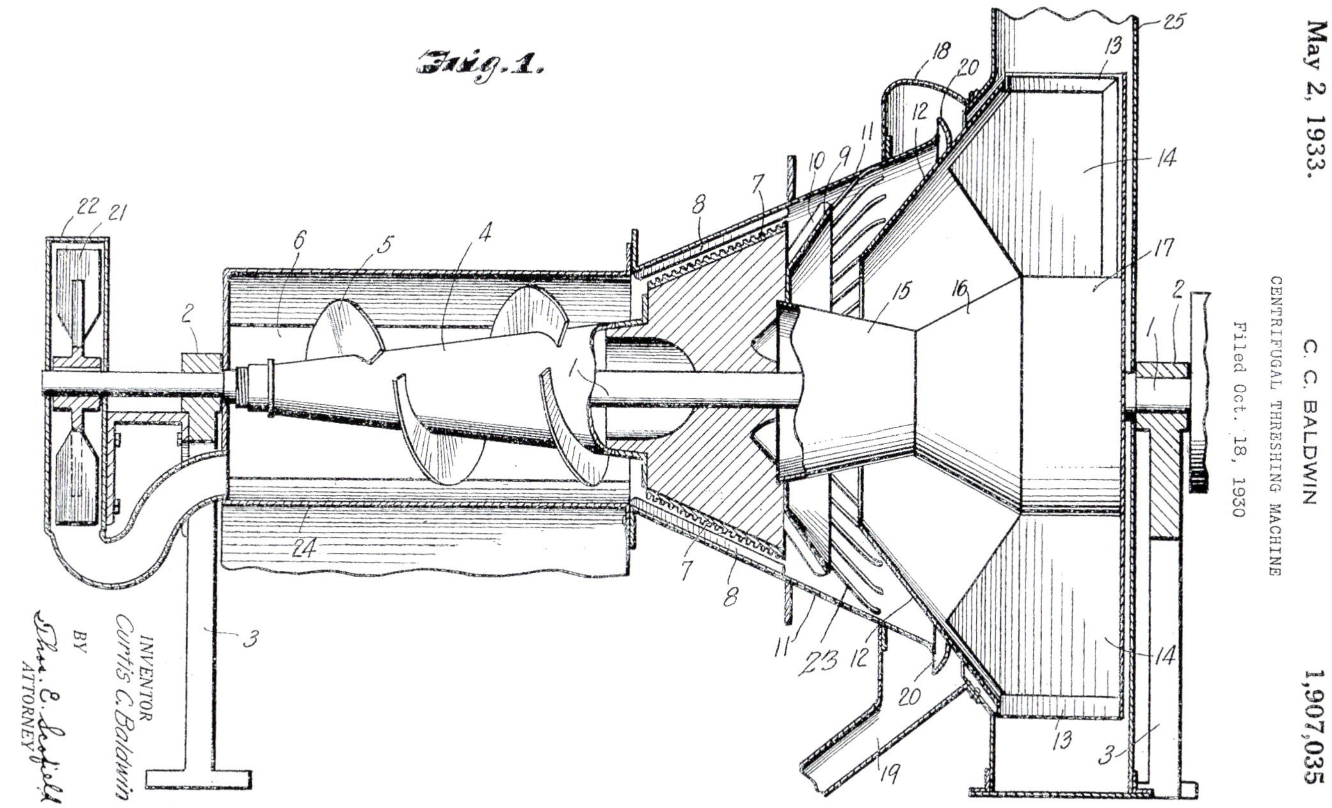

Centrifugal Threshing Machine

◀ Curtis Baldwin's design—as shown in this May 2, 1933, patent—used centrifugal force and air blown through the cylinder to separate grain. *U.S. Patent Office*

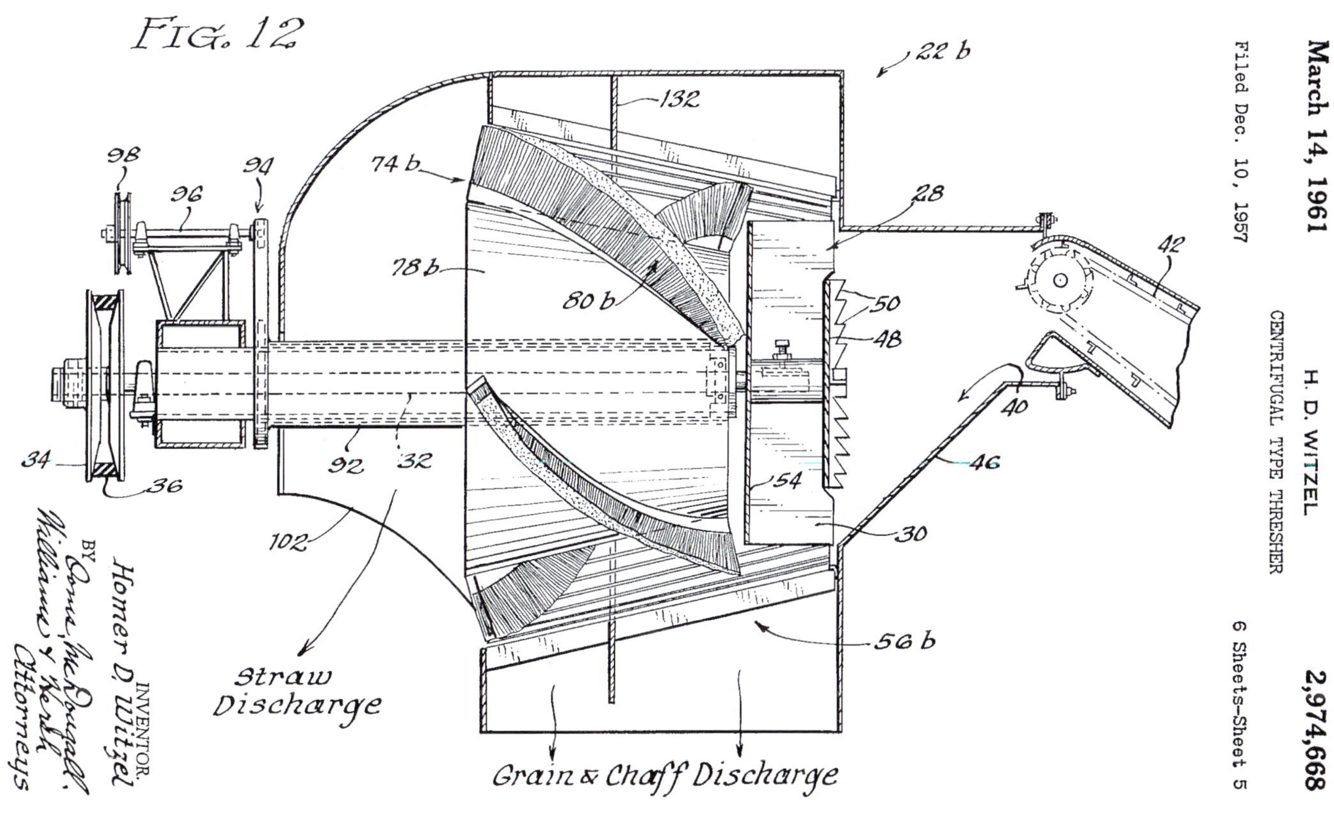

Harvestaire Design

◀ This patent filed in 1957 states the design is used to thresh grain and separate the grain kernels. Advantages cited include the ability to separate regardless of slope or angle—meaning it was ideally suited for hillside harvesting—and less cracking of grains. In 1957, the advantages of a rotary design were well-documented—the issue was that no one was able to build a machine that worked. Harvestaire tried mightily to make the design a reality, using a system that relied on forced air more heavily than later, more successful systems. Harvestaire built working prototypes but ultimately failed. *U.S. Patent Office*

HARVESTER'S ROTARY SEPARATION EXPERIMENTS

The Harvestaire machine attracted the attention of many in the industry, including the Harvester engineering team. Van Buskirk tested the machine in 1955 and wrote a report on how the machine compared to the International Harvester Model 141 combine.

The study of the Harvestaire in 1955 was part of a strong effort by Harvester to understand rotary threshing. Harvester had done extensive lab tests of conventional combines in the early 1950s. A 1977 ASAE paper by DePauw, Francis, and Snyder states: "The results of these studies concluded that the concave and beater were much more efficient than the straw walkers in separating grain."

Van Buskirk and Karlsson understood the concepts well. Beginning in 1955, they built a number of experimental separators in an effort to use a mix of centrifugal force and air to remove grain. Karlsson designed and built a cyclone chaff separator in April 1955, and Van Buskirk followed up a year later with a radically different rotary separating unit that

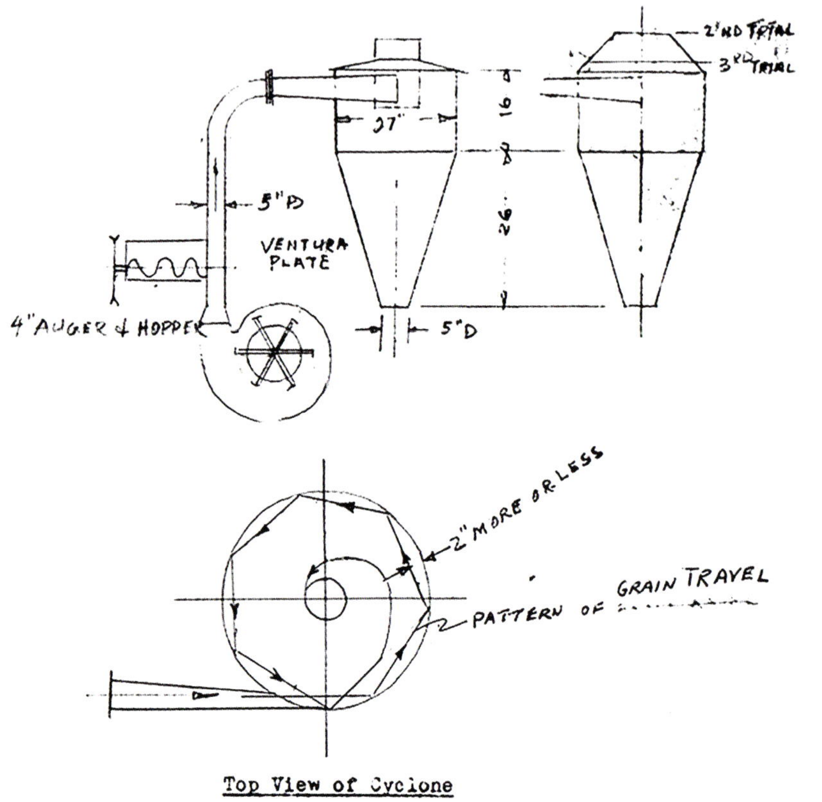

April 1955 Cyclone Chaff Separator

▲ Elof Karlsson designed this experimental unit to separate chaff from straw. After adjusting the air velocity, the shape of the cover, and the size of the cylinder, as well as applying a coating on the inside of the cylinder, he was able to successfully separate wheat. The losses and cracking were unacceptable, however, and Karlsson concluded that a more effective design would use a lower air velocity and a longer, narrower cylinder. *East Moline Works Report Case IH*

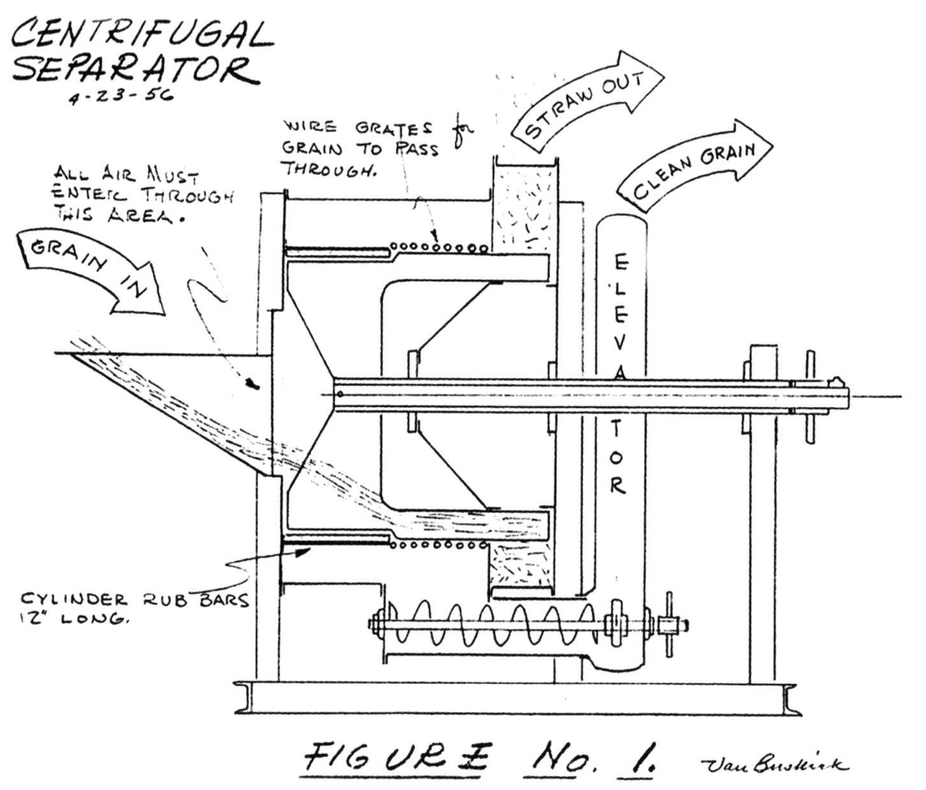

April 1956 Rotary Separation Unit

▲ This Mel Van Buskirk sketch is from a report about a rotary separation unit that was built and tested in April 1956. Initial runs went poorly. The airflow blew the straw and grain through the cylinder with no separation at all! On the third redesign of the unit, Van Buskirk reversed the rotor and was able to tune the stationary test machine to properly output about half of the grain. "Additional study of the problem will be made," he wrote. *East Moline Works Report Case IH* applied the same principles, but did so in a more compact design that used a very short 12-inch rotor.

In November 1956, Van Buskirk designed and tested a rotary separator that was 24 inches in diameter and 48 inches long. This longer, larger rotor design was the first of the early designs to clearly resemble the Axial-Flow combine as it is known in modern times.

In the first test of the new design, the grain did not even move across the drum. Performance improved marginally with radical changes to the design, but the report closes with Van Buskirk admitting the results would "not be considered satisfactory."

As the holiday season closed out 1956, the Axial-Flow combine was nothing but a badly performing, cobbled together idea stuffed in a closet. The only people who believed in the machine appeared to be two engineers with vision and a shared passion for dreaming up new technology.

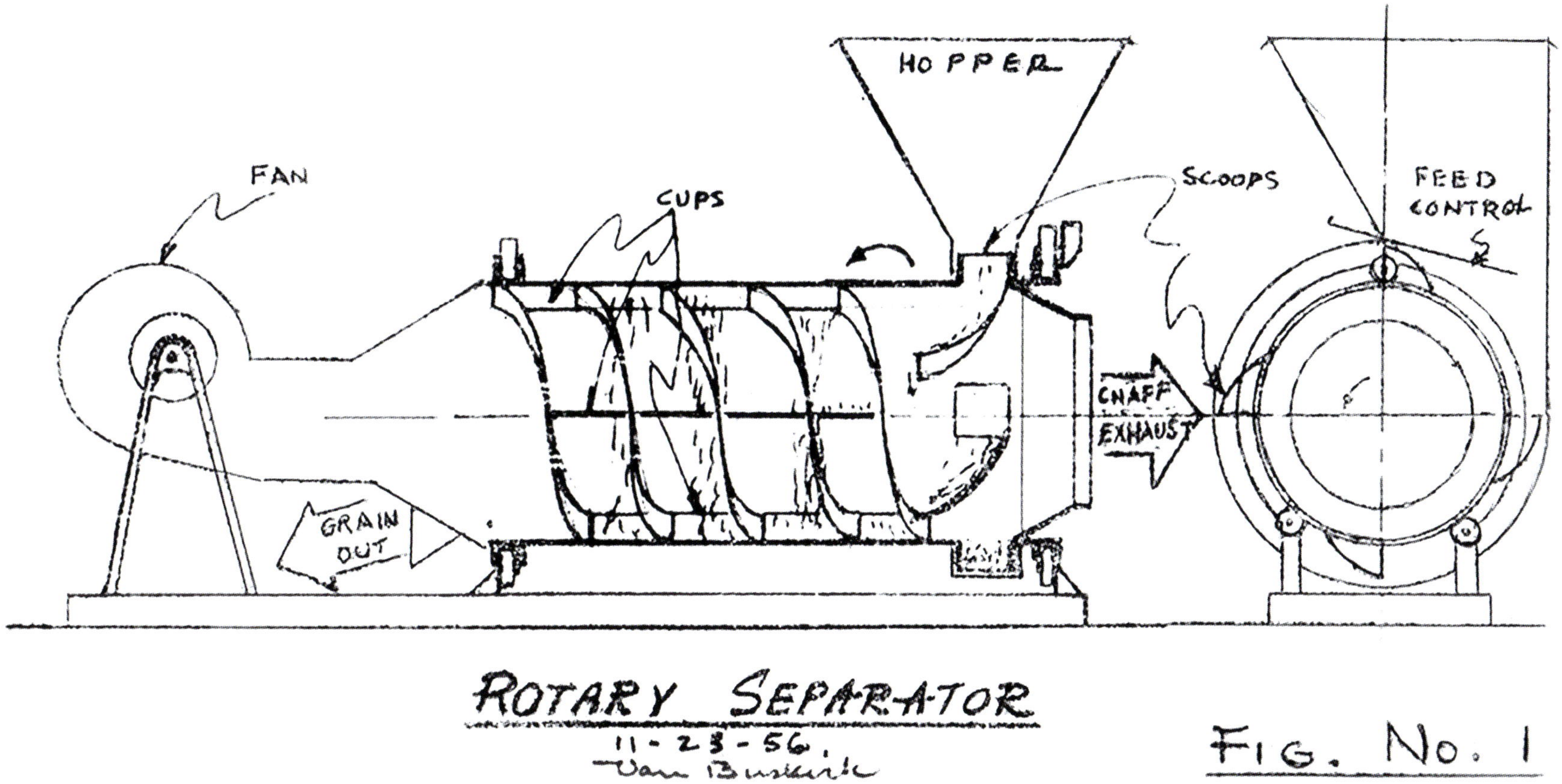

May 1956 Rotary Design

▲ This Van Buskirk sketch of a rotary separator was created in May 1956. A crude prototype was assembled in July 1956. The auger was 24 inches in diameter and turned by an electric motor at 21 rpm. A corn-picker fan driven by a Briggs & Stratton motor pushed air through the cylinder. In the first tests, the grain didn't even move through the cylinder. More testing took place in November 1956, when 100 pounds of material was fed into the machine in 30 seconds. The grain discharge was dirty, and it took about two minutes to run through. "The grain sample was cleaner than earlier tests," the report stated, "but would not be considered satisfactory. Plans for a new drum are under consideration."

East Moline Works Report Case IH

THE IH ROTARY IS BORN

Karlsson and Van Buskirk also installed their rotary separation cylinder into a full-sized combine for testing. A number 76 Harvester Thresher using a second cylinder installed behind a conventional cylinder was tested in late 1956 in California and South Dakota. Van Buskirk reported acceptable separation but an abundance of trash in the grain.

Ken Johnson remembered working with the Model 76 experimental combine. "It was built from a 76 combine, a small pull-type combine," he said. "It had a wide cylinder. It had a conventional cylinder for threshing. Behind that, it had another cylinder, and that cylinder was to take the place of the straw racks.

"That was the first rotary machine. It had some kind of a rotary cleaning system also, but needless to say, it wasn't very successful in the field."

By March 1957, the plans were laid to build another experimental device that was known as No. 10. According to Don Murray's memoirs, the machine used a conventional threshing cylinder that fed a rotary separator.

A prototype of the No. 10 was built in 1956 and tested in Oxnard, California, in late June. The machine design was improved, and the rebuilt unit was tested in Twin Falls, Idaho,

No. 10 Experimental Combine

▶ The No. 10 was the second experimental rotary combine. Harvester tested this machine from 1955 to 1958. The threshing cylinder was conventional, but the separator was a rotary, three-cylinder design created by Mel Van Buskirk. The rotors were arranged crosswise in the machine. The project was killed in 1958 when Stuart Pool took over Advanced Engineering and decided rotary separation technology wasn't worth the investment. *Dave Gustafson/Case IH*

in August and September 1957. Van Buskirk reported that the cleaning units did not have enough capacity.

By June 1958, the team had built a second No. 10, and the field results were frustrating. Not only did the cleaning drum perform poorly, but the machine experienced drive problems as well. To make matters worse, cylinder performance was worse than the previous year.

Another prototype, the No. 7 Rotary Combine, was designed in late 1958 and tested at Hermosillo, Mexico, in March 1959. The results were so dismal that the machine was discarded and the team's four years of research scrapped along with it. In October, another rotary project Van Buskirk was working on, the four-row self-propelled harvester, was also discontinued.

Developing a new combine is an amazingly difficult process, a point repeatedly stressed by the engineers interviewed for this book. The problem is that combines work in an incredibly dynamic environment with conditions that are almost infinitely variable. Crop moisture content varies, as does the amount of grain or corn on the plant. As a result, the materials going into the machine are not uniform—they vary widely. An oats field in Australia is typically quite different than one in Saskatchewan, yet the same machine has to harvest those plants.

In addition, there's the variable of weather. Add a fresh quarter inch of rainfall to a field, and harvesting the crops is an entirely new ballgame.

What this means is combines require extensive testing in a wide variety of conditions and environments to be properly developed. Combine test engineers routinely travel across North America to test machines. Tests are also commonly performed in Australia, South America, and Europe.

When a new design is created, even a modest upgrade of an existing machine, minute changes can reduce performance. Any time a design is altered even slightly, several seasons of testing are required to perfect the design.

FIG.1

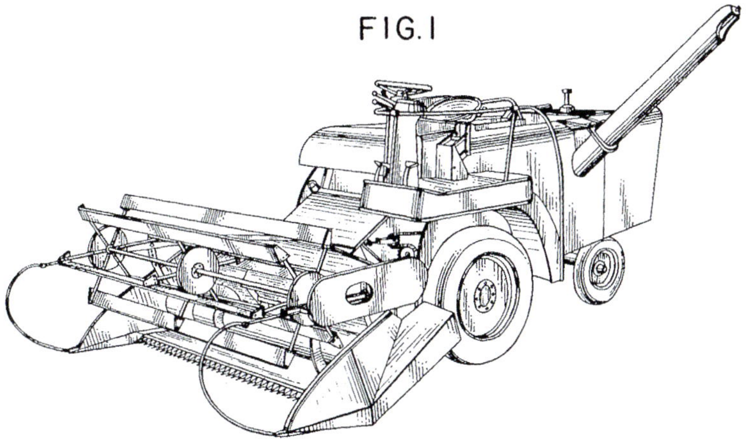

Combine Harvester and Thresher

▲ According to Don Murray, these drawings by Mel Van Buskirk for U.S. Patent 188,955 depict the No. 10 combine. *U.S. Patent Office*

FIG.2

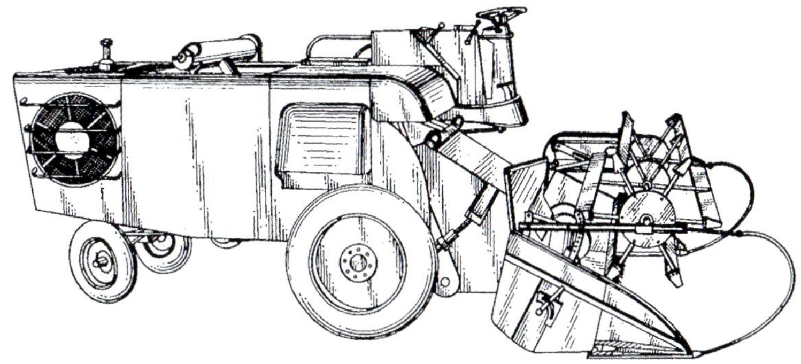

Patent Sketch #2

▲ This drawing's resemblance to the photograph of the No. 10 is clear. Van Buskirk's patent related mainly to the appearance of the machine. According to some Harvester records, a second No. 10 was built and tested in June 1958. *U.S. Patent Office*

For any combine development team, the testing process can be long and frustrating. For a team developing a completely new technology, the process can be utterly baffling.

A critical mass of sorts developed in the 1960s, with a variety of companies doing research and development of combines that used some kind of rotating device to separate the grain. John Deere, Massey-Ferguson, and New Holland all developed the technology during the decade. Axial-Flow technology was tested in Asia, France, Australia, and Canada as well.

The U.S.-based Harvester team was working in parallel with the IH France engineers, who were developing a rotary cleaning system.

The experience of the Harvester team—investing years of research and development into a concept in the late 1950s, only to have it all end up on the scrap heap—was not unique. In fact, it was the norm. In the 1960s, at least, no one was able to put the pieces together and create a commercially successful machine.

Development in the late 1950s hadn't resulted in a single product that had even a glimmer of hope of being a commercial success. But it had planted seeds in some very bright minds.

No. 7 Experimental Combine

▲ This experimental combine used a rotary cylinder mounted longitudinally. It was photographed on March 10, 1959. *Dave Gustafson/Case IH*

INSPIRED BY CORN

On November 8, 1960, Elof Karlsson and Mel Van Buskirk visited a Harvester dealership in Quincy, Illinois. Van Buskirk's notebook states that the men sat in the office of Mr. Selby of the IH dealership, Selby Equipment. What was said is lost to history. What resulted would make it.

At the time, combines used to pick corn were raising hell with straw walkers. They would tear up the system, and the result was a frustrating string of failures. American farmers were growing corn in increasing amounts, and the demand for corn combines was up. The technology, however, was inadequate.

In Mr. Selby's office, one can imagine the three men debating this thorny problem and the two engineers realizing that the rotary technology they had been testing during the past five years could solve the corn harvesting problem. In that room, the three men concocted the idea to use a rotating cylinder in a corn picker and sheller. The spiral system was simpler and more robust. The machine that resulted would be known as the 2MX Picker-Sheller.

By January 1961, construction of a test unit was approved. In April, a mockup of the 2MX was built at Hinsdale using a Model 141 chassis. The machine was hidden in a dairy barn "for security purposes."

Secrecy was a difficult proposition for combine developers. For one thing, the machines are huge, making them very hard to conceal during testing. Plus, the test sites favored by most

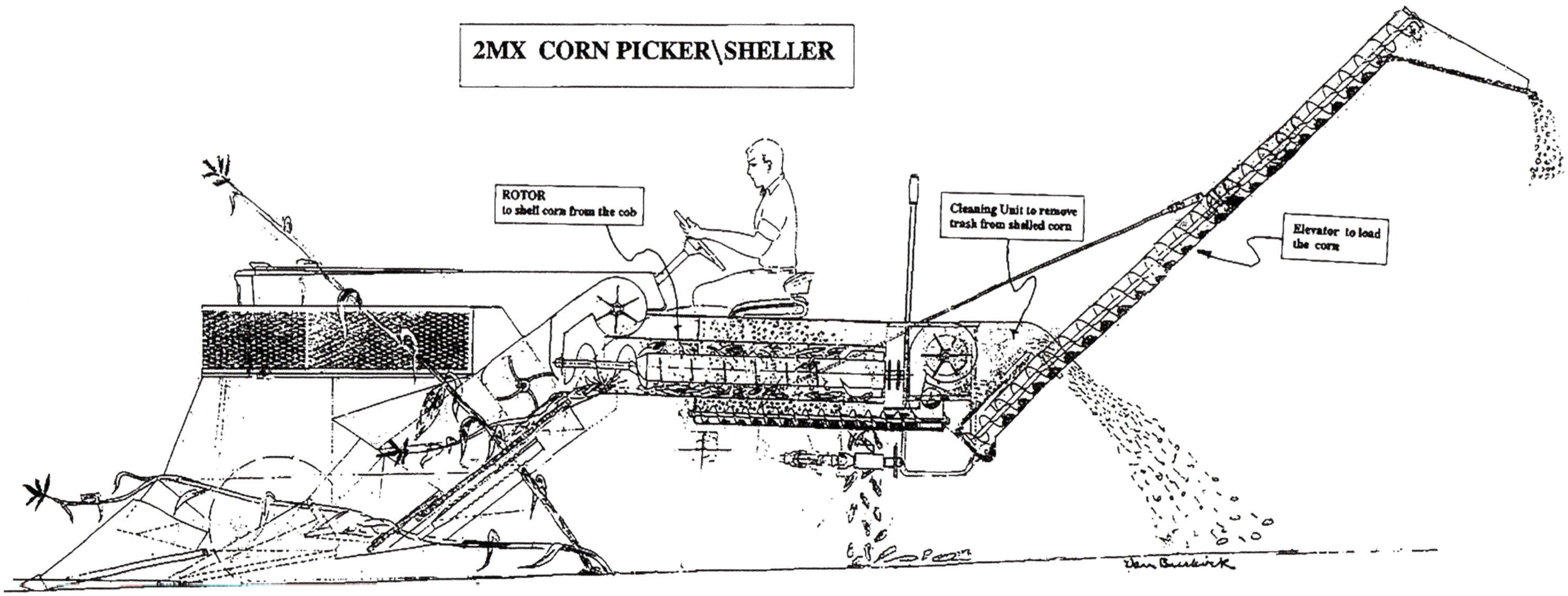

2MX Corn Picker/Sheller

▲ Sketched by Mel Van Buskirk, this machine was designed to use rotary separation to shell corn in the field. The resulting experimental machines were built inside a secret garage on the grounds of the combine plant in East Moline, Illinois. The machines were tested from 1960 to 1962 and performed well in the field. *Case IH*

Harvester Dealership circa 1955

▲ The idea for a rotary corn picker/sheller was born in a Harvester dealership in Quincy, Illinois, on November 8, 1960. *Wisconsin Historical Society #58848* engineering teams are fairly common. Stories of engineers finding competitors' machines sitting out in the field and inspecting them are common. Engineers have been known to go so far as test drive and disassemble what they found.

Not to mention the fact that John Deere's combine engineering and manufacturing was located in East Moline—the Harvester and Deere buildings practically shared a lot.

The 2MX, however, was developed in complete secrecy inside a garage at the East Moline Plant. That garage was a nondescript cement-block building with frosted windows. Access to the garage for the next 17 years would be strictly controlled. Only the key engineers working on the project were allowed inside. This unassuming garage would become one of the the most important buildings in the vast Harvester empire.

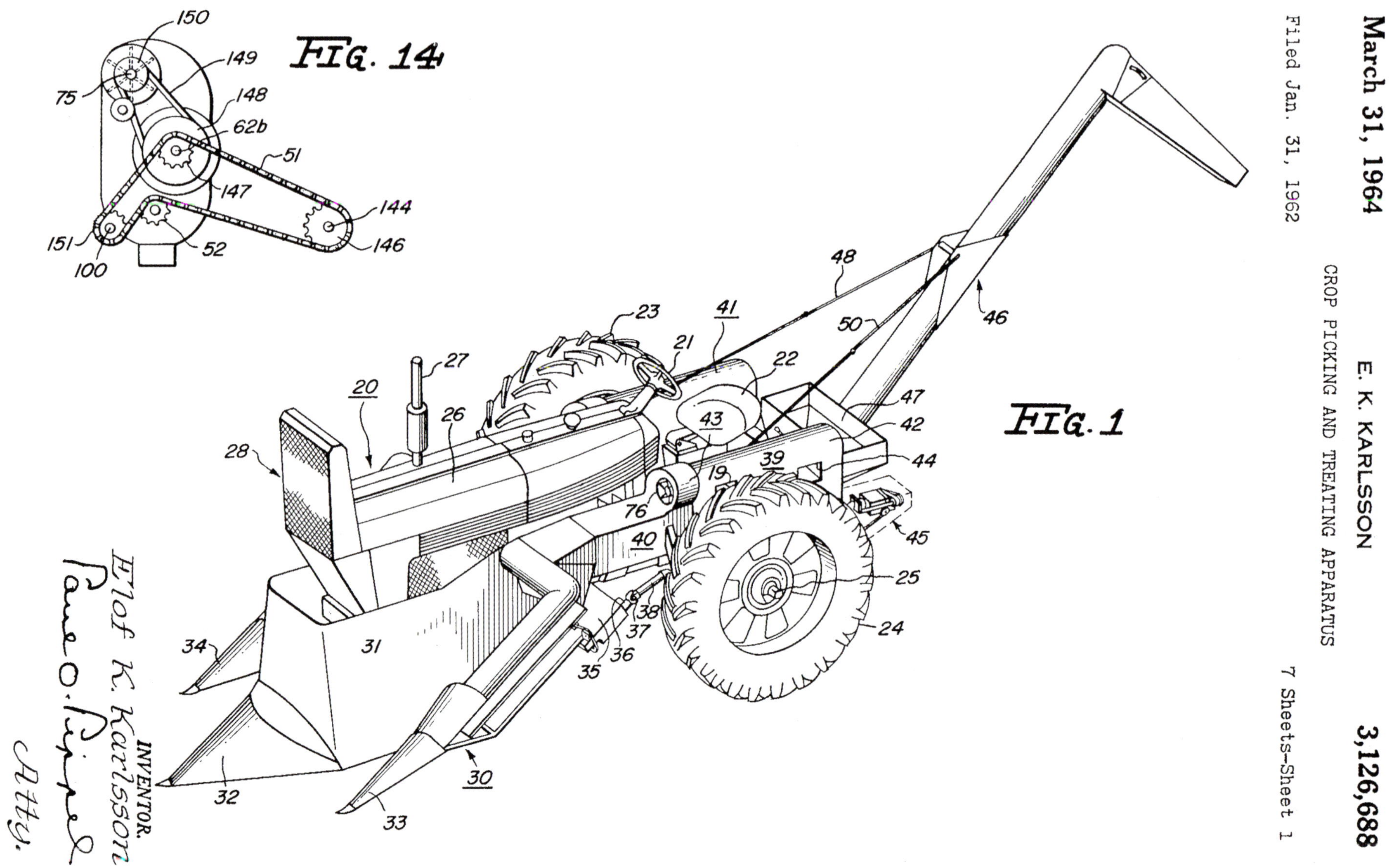

Elof Karlsson Corn Picker/Sheller Patent

▲ In January 1962, Karlsson filed a patent for the picker/sheller system, which was intended to provide better weight distribution and improved conservation of corn. The 2MX project was scrapped before year's end, however, presumably because of excessive development costs. *U.S. Patent Office*

The key engineers on the 2MX project were Elof Karlsson and Mel Van Buskirk. Harvester engineer Dick DePauw joined early in the 2MX development. The three were the core developers on the project and were supervised by Bill Adams and Rey Barkstrom.

The 2MX was delivered to Roseville by truck on September 29, 1960, and operated for the first time on October 2, 1960. Testing took place over the next few years, and the machine performed reasonably well. The 2MX project was shelved, however, late in 1962. The most likely reason was because the corn sheller Harvester had on the market was lower cost.

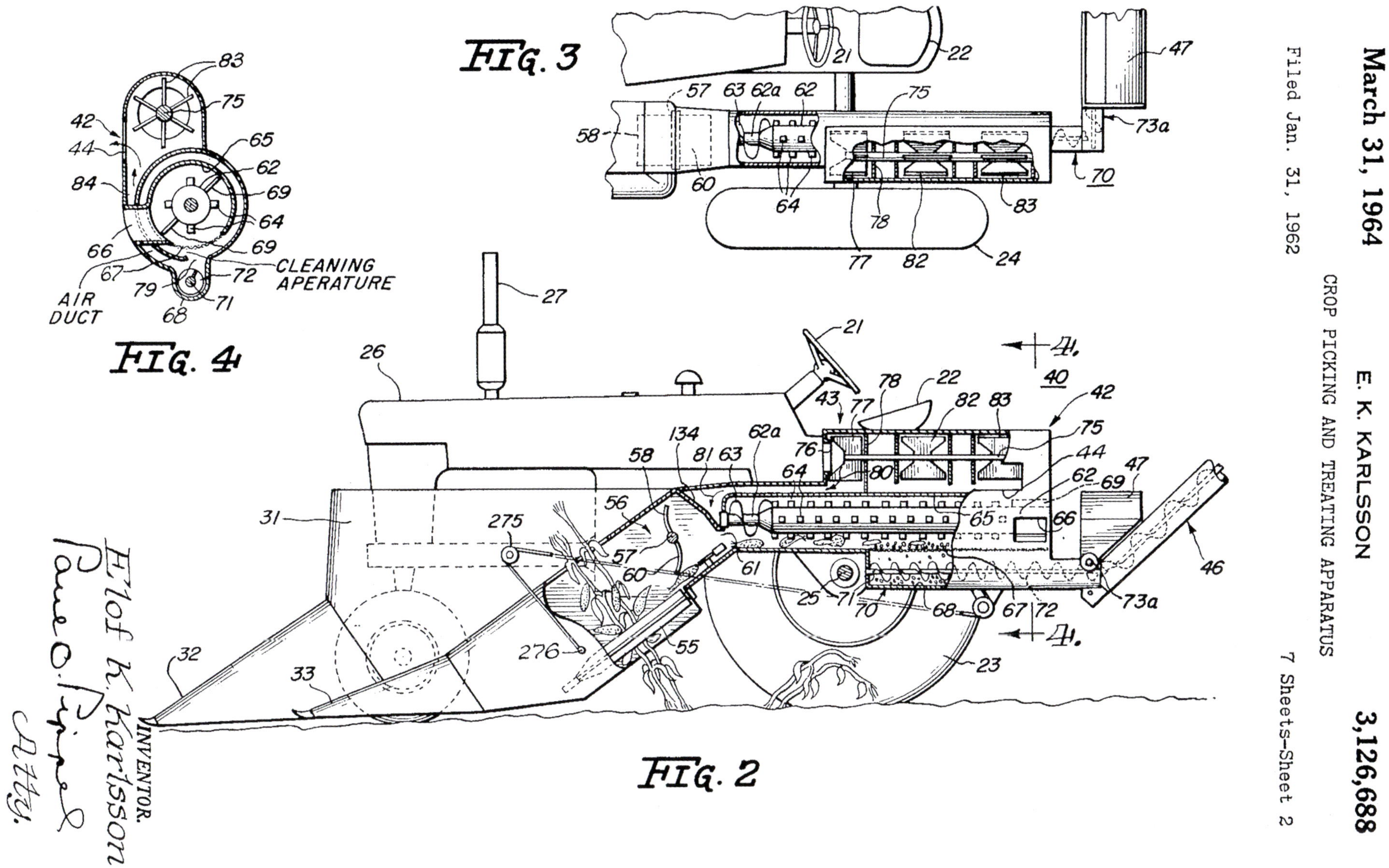

Patent Drawing Show Separation Process

▲ This drawing from Karlsson's patent diagrams the rotary separation of the corn picker/sheller.

U.S. Patent Office

THE MULTI-PASS COMBINE

The demise of the 2MX didn't deter Van Buskirk and Karlsson. In fact, it inspired them.

While testing the corn sheller, the pair had discussed at length how they might make the rotary combine idea work. In a letter, Van Buskirk wrote: "We favored a plan that we had discussed at some length: Why not take a 141 combine and put one large rotor where the cylinders and the straw racks were located and use a conventional cleaning system in its normal position?"

The men did just that, and by January 1963 a plan was in place to exhaustively test various rotors and concaves to create a new rotary combine dubbed the "multi-pass combine." Karlsson and Van Buskirk worked furiously in the secret garage, running tons of straw through different types of rotors and concaves. The men worked harder than ever before to make the rotary concept a reality.

The secrecy of the multi-pass combine was considered of paramount importance. On January 31, 1963, a letter was issued stressing the security surrounding its development.

Don Murray was the chief engineer at Harvester in the mid-1960s and worked at Hinsdale before later being transferred to East Moline.

"The system was so secret there really wasn't much outward mood," he said about the multi-pass combine's early development. "There were probably only eight or ten people altogether that even knew what was going on. The parts were made in our experimental shop and then the assembly was all

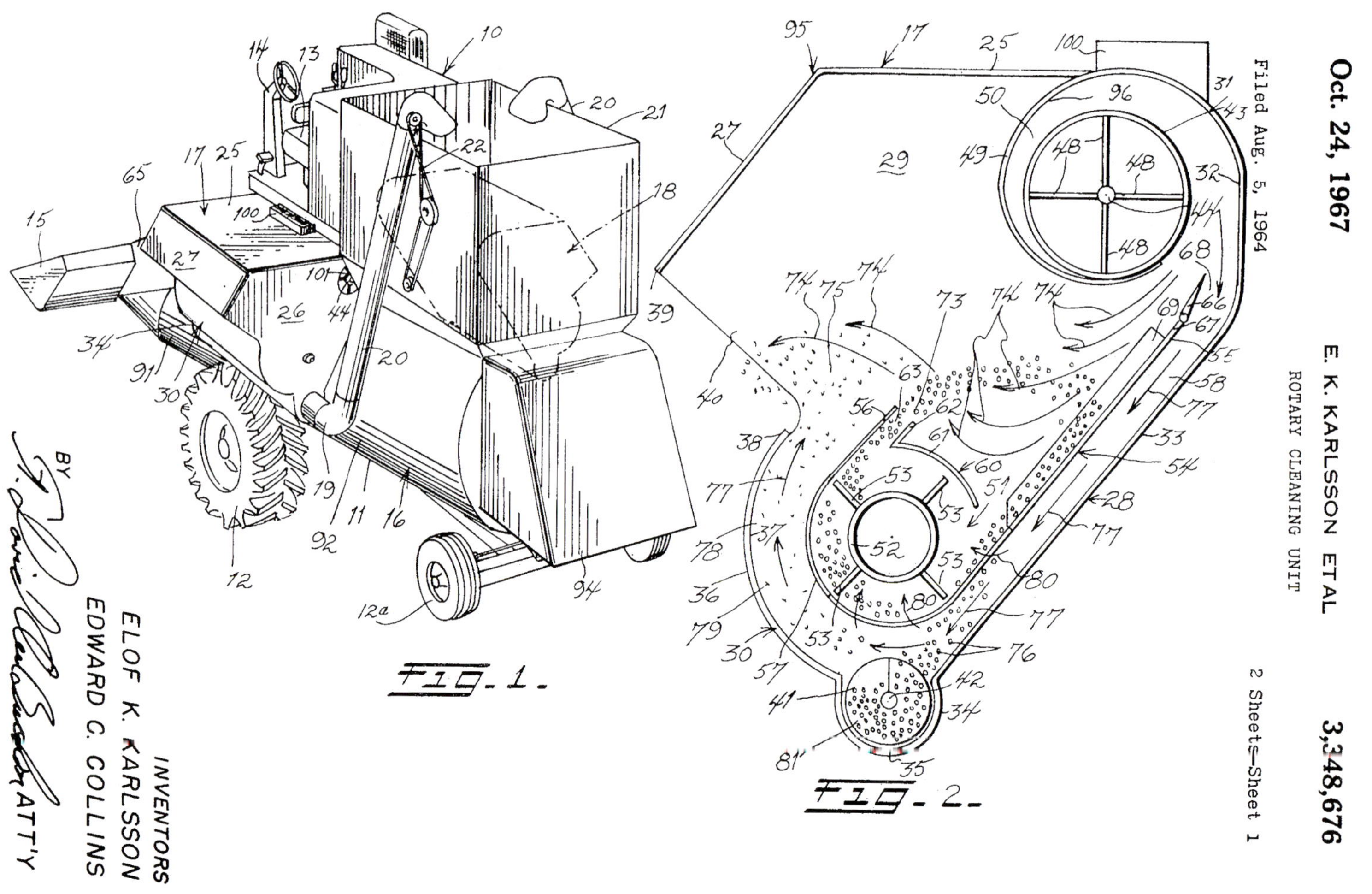

Rotary Cleaning Unit Patent

▶ In the mid-1960s, Harvester went back to work on the rotary combine concept. This Elof Karlsson patent cites many of the advantages of the design: improved efficiency and fewer parts than the straw walker design that dominated the market at that time. "Not until just recently," he wrote in the patent application, "has any appreciable attention been given to the so-called axial flow combines." That was changing rapidly in the mid-1960s.

U.S. Patent Office

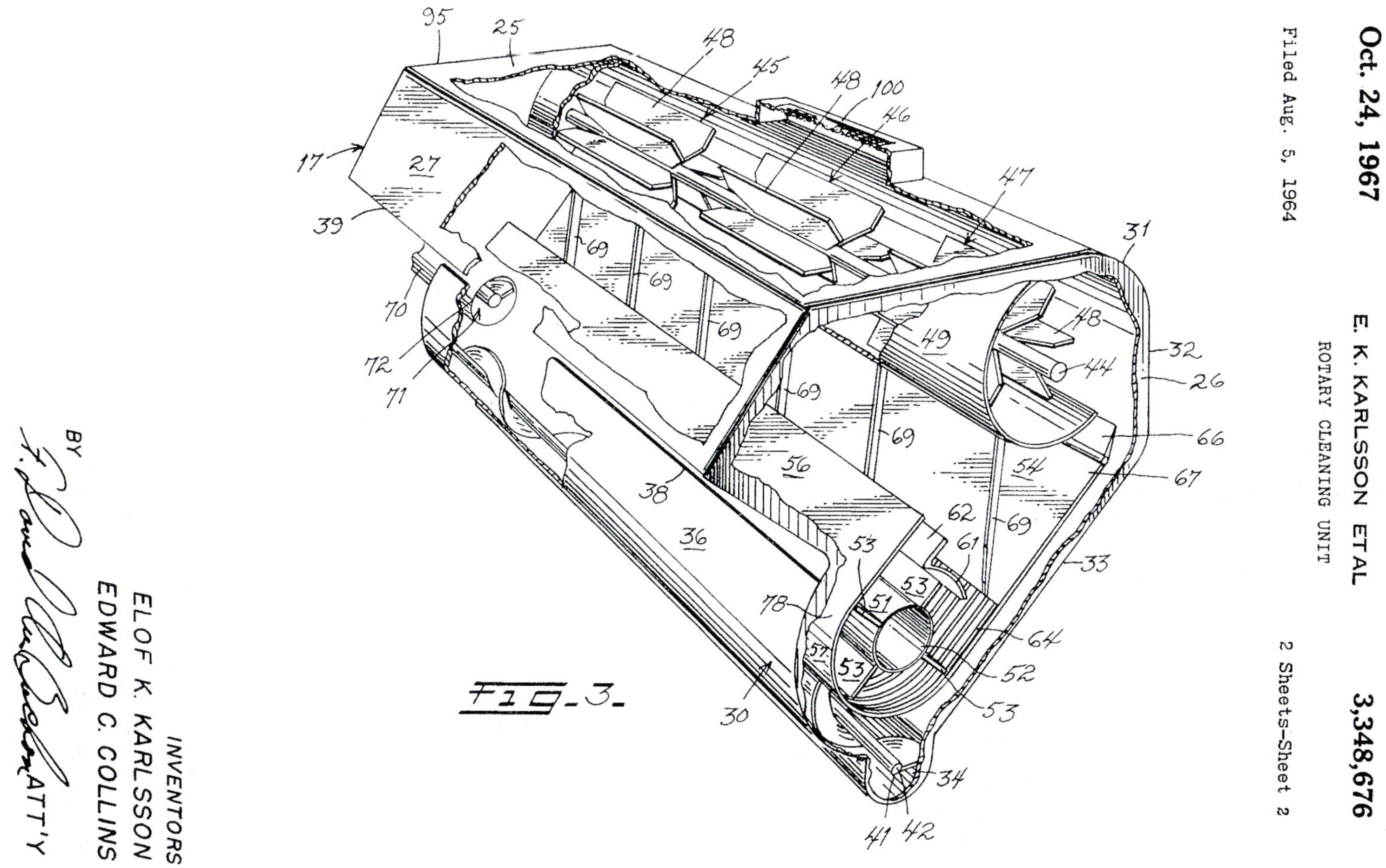

Rotary Cleaning Unit

◀ This patent drawing shows the rotary cleaning unit. The patent was applied for on August 5, 1964. *U.S. Patent Office*

done by one or two men, and maybe an engineer or designer in what we called the garage, which was a part of the engineering experimental shop. . . . The governing product committee didn't know [the multi-pass combine] existed, and it was listed on the engineering budget as a miscellaneous item."

The secret garage would become a second home for Karlsson, Van Buskirk, and DePauw, who worked tirelessly on the project no one else knew about.

By May 1963, Van Buskirk reported that the rotary unit had a higher capacity than the Model 503, but that the cleaning system couldn't adequately handle the output and required further development. In June, dual 303 cleaning systems were fitted, and by July it was determined this could adequately handle the output of the rotor.

The concave was lengthened in July 1963. After months of testing, Van Buskirk reported in late September that the longer concave was a step backward. The cleaning units were lowered, a return pan was installed, and a straw ejection auger built. In November, the overloaded team reported it needed more help to keep the project on schedule.

In December, they reported that the cleaning system was still inadequate. By January 1964, the decision was made to use a slightly smaller 24-inch rotor and install it in a 403 combine rather than a 503 as originally planned. That machine was specified, designed, and ready to be painted in April 1964, but the paint was applied poorly, which caused more delays.

THE CX-1

By June 1964, the team had the first multi-pass combine ready for testing. The engineering team decided to name the prototype "CX." The project (later the Axial-Flow) was the first CX, so the model was the CX-1.

It was shipped to Tucson, Arizona, and unloaded west of Casa Grande.

This would be the first major test of the new CX-1, and it would be attended by most key members of the team. Karlsson and Van Buskirk were there, as was Bill Adams, the assistant chief engineer, and Frank Roberts, the mechanic who assembled the machines in the garage. Rey Barkstrom—who led the project in conjunction with Adams—was back in the office in East Moline.

The entire visit was kept a secret. All the men on the test were forbidden to tell anyone—even family members—where they were going and what they were doing.

Adams made arrangements to rent a large, private ranch equipped with a lockable storage shed. The CX-1 would be kept under lock and key, hidden from prying eyes. The ranch

The CX-1 Experimental Combine

▶ The next generation of Axial-Flow experimental combines was the CX-1—a 24-inch rotor installed in a 403 combine shell. Most of the components came from a 403. The first test unit was sent to the field in Arizona in early 1964. This image is most likely from 1965.

Dave Gustafson/Case IH

The Francisco Grande Hotel

◀ Early testing of the CX-1 was done in Casa Grande, Arizona. The Harvester engineering test team stayed at the Francisco Grande, a hotel built to house the San Francisco Giants and favored by celebrities, including John Wayne. *Jordan Smith*

was a working cotton farm not far from Casa Grande owned by actor John Wayne.

The Harvester test group stayed at the Francisco Grande Hotel, which was built in 1959 to house the San Francisco Giants baseball team during spring training. The pool was shaped like a baseball bat, with the hot tub evoking a baseball. A giant statue of a baseball bat sat outside the hotel.

The swanky Francisco Grande became a bit of a haven for celebrities, including Wayne, who, legend has it, liked the place because they let him target shoot off his room's balcony. While there is no record of movie star encounters while the Harvester engineers were at the hotel, the group would discover they had interesting company.

The CX-1 arrived in Tucson and was transported to the ranch. The first test was completed on June 2, 1963. For a few days the CX-1 was run in 90-bushel barley all day, and the team enjoyed the hotel each night. Shortly after getting settled in, the engineers discovered the Francisco Grande's notable guests: a group of Massey-Harris engineers.

The team continued testing despite the concerns about its rivals. They kept the CX-1 under cover and, while it was being tested, maintained a watchful eye on the surrounding fields.

On the third day in the field, a car drove out to meet them.

"I'd made arrangements with my mechanic that if I give him the high sign that we'd clear out," Van Buskirk said. "I knew Massey-Harris engineers were in the hotel with us."

Van Buskirk's concerns turned out to be valid: the car contained some of the Massey engineers. Several of them got out of the car and asked for Bill Hill, a Harvester engineer at the time.

Pool at the Francisco Grande Hotel

▲ The Francisco Grande was the site of the San Francisco Giants' spring training facility. The pool even had a view of the ballfield. While staying here for early CX-1 testing, the Harvester team discovered the hotel also housed combine engineers from a rival manufacturer. *Jordan Smith*

"I explained we did not want any visitors," Van Buskirk wrote in his journal.

The CX-1 was secured under lock and key. Van Buskirk called the chief engineer, who instructed him to get on a plane to Colorado immediately to scout a new location to test. The harvest was nearly done in Colorado, so the team eventually decided to move testing to Pendleton, Oregon.

On June 26, the CX-1 was transferred from John Wayne's cotton farm to Temple Ranch near Pendleton, Oregon.

"We ran our second test up there and we never had any problems at Pendleton and that area with security," Van Buskirk said.

The testing proved promising as well as frustrating. They had to wait for the crop in Oregon to be ready. Once it was, the machine showed great promise at times but also had all the teething issues one might expect of a raw, hand-assembled prototype.

During the test at Wayne's ranch, Van Buskirk had to drive Bill Adams to the Tucson airport. Adams had been promoted to chief engineer of Stockton Works, where his presence was required.

When Adams left, Rey Backstrom placed one of his new guys, Edward William Rowland-Hill—or simply Bill Hill, as the engineering team called him—as product engineer on the CX-1.

In fall 1964, around the time when Hill began, Elof Karlsson retired at the age of 62. Don Murray wrote of the event, "We don't know whether he retired because of having been taken off the project or whether he was off the project by the simple fact he retired."

Hill, in the meantime, was doing his best to get up to speed on the CX project. He wrote a review of the basic design that was dated September 9, 1964.

According to Murray, however, Hill wasn't allowed to go in the field and even see the CX-1 until November 16. Dick DePauw reported that late in 1964 and early in 1965, he and Mel "taught Bill all we knew about the rotor."

Murray reported that Backstrom and Hill had a big "shootout" and Hill quit Harvester in early 1965. According to Van Buskirk's log sheet, Hill's last day at International Harvester was February 23, 1965.

This wouldn't be the last that the combine boys would hear about Bill Hill. Not by a long shot.

NEW BLOOD

Donald Murray started at East Moline after being chief engineer at two other plants. Before coming to Moline, he was the chief engineer first at Stockton and then Hinsdale.

He first learned of the CX-1 project while training Bill Adams, his replacement at Stockton. Adams had been leading the team and told Murray about the "super secret project" he was leading.

Murray was later transferred to Hinsdale, then to East Moline. Going into East Moline, he knew something special

Chief Engineer Don Murray

▶ Don Murray was transferred to East Moline on September 1, 1966, to be the chief engineer for Grain Harvesting Engineering. He was the leader of the Axial-Flow project, and his detailed written account of the process provides an insider's view of the trials and tribulations faced during the development of the Axial-Flow combine. *Case IH*

was happening. His first opportunity to see an Axial-Flow machine in the field came when it was being tested in Arkansas rice. He was impressed first by the lack of fear displayed by his new engineering team, which was willing to test the machine in the toughest of conditions.

"Boy, talk about guts," he wrote. "The toughest crop we regularly harvested and there we were."

Performance at the test was mixed. It did moderately well—enough to give Murray "a false sense of confidence." What he meant by that was thousands of hours of development and millions of dollars would be required to ready this impressive machine for production. This was 1966.

Even Murray's tempered predictions would be a dollar short and years late—the team would need another 11 years to get the machine to market.

Murray's first exposure to the CX-1 in the field left him impressed by the security surrounding the project. Parts were built in the engineering experimental shop at East Moline and then taken into the secret garage for assembly. "Only a half-dozen men from the whole outfit [were] allowed in," he wrote. "It was shipped out either under a tarp or in the dark of the night or both."

The truck driver was not to open his destination orders until he got to a certain intermediate location where he was met by a company contact. The regular test crews never operated the CX-1, and it always ran alone or with only one other machine. And the place where it was tested was off the beaten path and featured a lockable building for storage.

In addition, the project wasn't even on the books. The engineering budget listed it as "miscellaneous." The Product Development Committee didn't even know what the machine was until the late 1960s. Prior to that time, they just heard rumors.

The CX-1 project was so secret that most of International Harvester Company wasn't even aware it existed.

THE NEXT GENERATION PROTOTYPES

According to Murray, CX-1 development slowed from 1967 to 1969 as the engineering group was focused on development of the new 815 and 915.

But testing continued. On June 29, 1967, the CX-1 harvested barley side-by-side with a Massey-Harris machine in Wasco, California. The CX-1 "compared favorably."

In July, the CX-1 rotor worked well in bluegrass, but cleaning was an issue. In September, the combine had feed problems with windrows in North Dakota. Later that month, it performed well in Illinois corn. In October and November, the CX-1 harvested soybeans well in England, Arkansas. In December, the machine was back in Illinois and working well in corn but suffering mechanical failures.

In 1968, CX-1 testing took place in barley in Merced, California, rice near Houston, Texas, and soybeans and rice near Little Rock, Arkansas.

Development may have slowed, but the CX-1 was getting around.

CX-18 and CX-14 Experimental Combines

▼ In 1969, the engineering team had completed its work on the new 15 Series machines and began working in earnest on the new Axial-Flow concept. Two experimental machines, the CX-18 and CX-14, were authorized in 1969 and first tested in 1970. Here they are seen working near Grand Forks, North Dakota, in August 1970. The CX-14 was the larger model and was built with a 30-inch rotor. *Dave Gustafson/Case IH*

Corn King

▲ This early CX-18 is harvesting corn near Roseville, Illinois, in March 1970. From the start, the experimental Axial-Flow combines worked well in corn. *Dave Gustafson/Case IH*

It was also gathering traction internally. On December 15, 1967, Don Murray filed an internal report on the machine recommending development of a corn-specific as well as a general purpose model, and he proposed building two new, larger experimentals: the CX-14 and the CX-18.

The report was very positive about the rotary combine. "We have had good acreage tests (over 1,000 acres in several crops) with the new CX-1 and rather clearly understand its shortcoming and strong points. . . . People who have worked with the machine (test, product, and farmers) are enthusiastic. . . . We have a good foundation to build on."

The report closed by stating that a corn-only model of the CX-1 could be in production as early as May 1970. Murray later characterized such projections as "the usual unsubstantiated enthusiasm."

Such optimism and enthusiasm were needed. Seven years would pass before the new technology was introduced to the public.

At some point in the late 1960s, product engineer William Knapp dubbed the Harvester experimental rotary machine the "multi-pass" combine. The engineers liked the simplicity and logic of the term—it referred to the fact that the incoming material made multiple passes through the rotary threshing and separating elements. The term caught on.

While the corn harvester idea would die a long, painful death, the plans to build the CX-14 and CX-18 proceeded.

The CX-14 was designed with a larger 30-inch rotor and two 403 cleaning systems. The main structural frame was larger than other Harvester combines, making it awkward and expensive. The cleaning bed continued to be a problem area for the development team. Harvester's existing technology used an oscillating grain pan, which worked poorly with the multi-pass combines. In 1970, John Deere came out with a new design that worked well and was patented, meaning the Harvester team's options were limited.

The Harvester engineers tested a belt conveyor system. Murray reported, "It was a disaster, both from a functional standpoint and a mechanical standpoint."

Work continued, and testing began on the CX-18 in 1969. Trends continued: the machine harvested corn like a demon, and struggled with long, tough-stalk small grain.

Murray marveled at the simplicity of the team goals at that point. The six-year planning report goal was "To build a corn harvester/combine with the durability and ruggedness of a corn sheller."

That was it. No discussions of separation efficiency, size, market—none of these were written concerns. Considering the fact that the Axial-Flow combine wasn't even a budget line item at the time, such simple goals are not terribly surprising.

The goals were unwritten, the budget was off the radar, and most of the company's gatekeepers and management didn't even know the idea for the Axial-Flow existed.

Murray, Karlsson, Van Buskirk, and a few other key IH people knew. They lived and breathed the project.

In 1969, that's all that really mattered.

Grain Concerns

◀ Experimental Axial-Flow combines struggled to work well in long, tough stalk grain. In 1970 and 1971, product and test engineering put all hands on deck to create a cone-shaped rotor front that vastly improved feeding.

Dave Gustafson/Case IH

The Best Bet

▲ At the time this prototype was tested in 1970, Harvester had determined that the Axial-Flow concept was the key to a competitive combine. Test results from harvesting more than 1,000 acres showed the machine to be an exceptional performer, and the feedback from the people allowed to use the machines was enthusiastic. The Harvester engineering team optimistically stated the machines could be ready for production in a year or two. They were right that the concept was revolutionary—they underestimated the massive development needed to bring the machines to market. *Dave Gustafson/Case IH*

THE GARAGE FILMS

Dave Gustafson started with International Harvester in March 1969 as a test engineer. That was the start of a lifetime with red combines.

"The first word I said was 'tractor,'" Gustafson said. "I was destined to be an engineer. . . . I grew up on Hs and Ms, so there was no doubt what color it was."

He studied agricultural engineering at Iowa State, and he worked with a group on a senior project to home-build a harvester that used a tractor chassis with a reversible operator's deck, two IH 234 shellers, and a four-row head. The group had unwittingly created a design similar to the CX-11 experimental. For Gustafson's design reviews, some of the engineers from East Moline came out. They saw his work, and when he was ready for a job, they hired him.

Gustafson has been working with red combines ever since.

On his first day at work back in 1969, he tested a 715 combine under the tutelage of Camiel Beert. Not long after that, he went to work for test engineer Ken Johnson to help with the top-secret project in the garage.

The garage was where Karlsson and Van Buskirk had started development of the rotary combine. In the early days, only a handful of people were allowed inside the building. When Gustafson started, roughly a dozen people worked there. Secrecy remained a priority. No one but the core group of engineers and mechanics were allowed inside. The windows were still frosted.

Behind the locked doors in the garage, the test setup for the rotary combine consisted of two stands. One contained the rotor, and the other contained the cleaning unit. Power was provided by an electric motor with a recording meter to capture input power data. The two units could be moved around the room to be run individually or paired to test separation and cleaning at the same time.

The test unit was fed with traditional bundles of grain and corn. A local farmer provided the bundles, which were stored in a barn in the area. In order to simulate wet crop, the material would be sprayed with a garden hose for a set number of seconds.

The crop was run through—Gustafson recalls 20 or 30 feet of it would run—and dumped out onto a large canvas for analysis. The material that came out was run through a recleaner and weighed to determine results.

The rotor and cleaning unit were shelled in Plexiglas so that crucial parts of the process were visible. Once a test was set up to run, word would go out to the engineers to come in and watch.

"Run after run, we'd run different feed rates and we'd run different designs," Gustafson said of the tests in the garage. "We would develop the parts in conjunction with other field studies we had. If the field had a problem, we would move it into the lab. If we came up with an improvement, we'd send it out in the field and say, 'It looks good in here. Does it look good out there?'"

The Secret Garage Tests

▼ The Axial-Flow combine was developed in this garage at East Moline Works. Access to the garage was carefully guarded—no one other than a select group of engineers was allowed access. High-speed films were made of the rotor at work. The movie camera is visible in the top left of the frame.

Dave Gustafson/Case IH

Late-Night Testing

▲ Material is run through the test head and rotor during Axial-Flow development. The grain was purchased from a local farmer and had to be carefully monitored so that the moisture level was consistent. The IH engineering team logged thousands of hours in this facility in the 1960s and 1970s to perfect their Axial-Flow combine. *Dave Gustafson/Case IH*

Flow

◀ This image from the test garage shows the Plexiglas sheeting that allowed engineers to film material passing through the concaves and into the rotor. The shape of the cone was tested exhaustively—the team made more than 600 runs in 1970 and 1971. "It didn't work one bit better on the last run," Dave Gustafson recalls. "But we learned a lot." *Dave Gustafson/Case IH*

Eyes have limitations, and rotors work rapidly. In order to accurately dissect each test run, the engineering team developed a unique system. Each run was filmed with 16mm high-speed film that recorded 400 frames per second. The film was taken to the local television station and processed the next morning along with the station's nightly sports footage. The film was picked up at 6 a.m.

The day after a test, the engineering team started their day obsessing over the film footage. "I often sat in on the first run," Murray wrote. "I was fascinated by what they could learn by studying that film, often a frame at a time in the special projector to see rate of flow and angle of movement of the material. Without that slow, stubborn process, I seriously doubt we would ever have had a really successful system."

Any hesitation they spotted on the footage meant trouble in the field. Hesitation meant grain was wrapping around the rotor or plugging the entrance or exit.

For the combine to function properly, flow of material had to be smooth. Finding that sweet spot was a painstaking process.

From the very beginning, the rotor worked wonderfully with corn and struggled with grain, particularly when the straw was large and tough.

"It did not like long strawed crops, especially tough crops," Gustafson recalled. "It didn't want to go in the front. When it got inside, it wouldn't go to the back. Then when it got to the back, it wouldn't go out the back. Every part of the rotor development was challenging."

The engineering group persevered. They had seen the machine at work in the fields, at times running faster and more productively than traditional combines from their own line as well as the competition.

FEEDING THE BEAST

In 1970, the garage test revealed several key breakthroughs. One was how the rotor was fed. The original design fed material into the rotor from below the rotor shaft. At some point in development, the feeding was switched to above the rotor.

Gustafson describes switching back to undershot feeding as a critical change and one of the key breakthroughs for the Axial-Flow. "In retrospect," he said of changing the feeding location. "It seems pretty simplistic."

Don Murray also mentioned the feed location in his memoirs, stating that he never cared for the high feeder location. "The header didn't float properly . . . it wanted to dig in and 'pole vault.' We had the same problem in the early 1950s with the Minneapolis-Moline rice combine."

A field report dated December 5, 1969, requested that the feeder be moved from the high position to a lower one. By June 25, 1970, the feeder was lowered and being tested in the lab. In January 1972, the first field reports from Illinois stated the lowered feeder was working well in dry corn.

The change improved feeding but in no way fixed the issues with straw flow. The material feeding into the tube continued to show those back-breaking hesitations in the lab that led to clogs and breakage in the field.

The problem area was the cone that funneled cut stalks into the rotor. The engineers identified that the problem was how it fed. The solution was to change the shapes of the cone and the impellers. The impellers acted as a large fan, blowing into the rotor and affecting the system all the

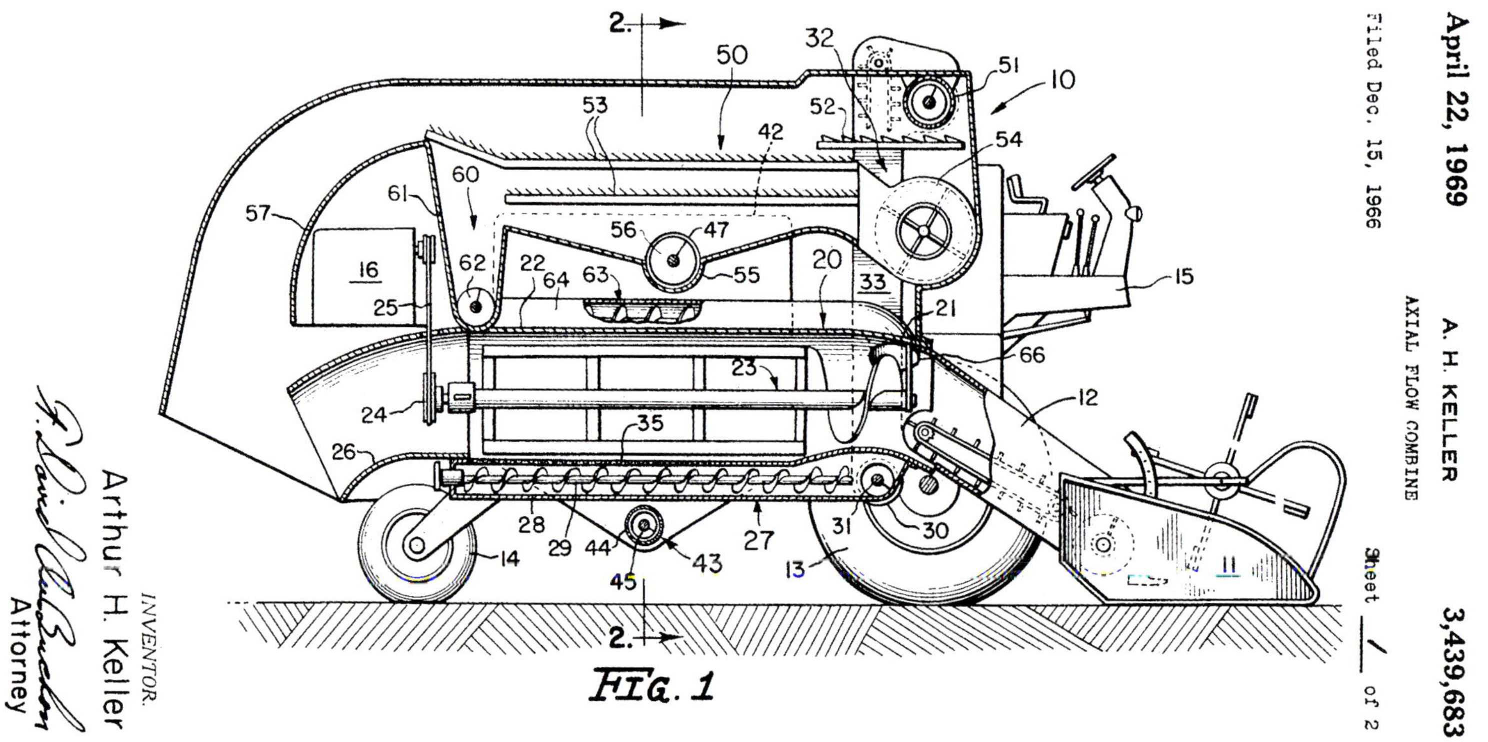

Undershot Feeder

▶ This patent drawing from 1969 features an undershot feeder, which is visible at numbers 12 and 30. The conveyor carries material up from the head and feeds it into the bottom of the rotor. For a time, the Harvester team used an overshot design, which fed material into the top side of the rotor. It switched to an undershot design in April 1972, and performance improved. *Dave Gustafson/Case IH* way to the back. The tough part was fine-tuning the size and shape of each.

Gustafson remembers a particularly frustrating stretch of cone and impeller testing that took place in late 1970 into early 1971. "We made hundreds of runs in the garage. It didn't work one bit better on the last run," he said. "But we learned a lot."

They eventually found a combination that worked, and in April 1972 test units were built with both the undershot feeder and the new cone-and-impeller feeding system. It was an improvement, but not a complete fix. Development would continue on this system.

Don Murray pointed out that the feeder system eliminated an age-old problem with conventional combines: dust being blown into the face of the operator. That was gone, but for the rotary combine, engineers had to manage air flowing out of the back of the rotor and raising havoc with the cleaning system.

THE REDESIGN

As if there wasn't enough to worry about, in 1971 a vice president's casual remark during a product review became gospel. As Murray recalled, it was division vice president Pat Kaine who suggested that the machine's design reflect the longitudinal rotor.

"We engaged the Industrial Design Department at Hinsdale," Murray wrote, "and some of their early suggestions were radical, more like a horizontal rocket than a practical combine."

During the winter of 1971–1972, the undershot feeder, refined impeller-and-cone feeding, and a new rear-discharge system were combined with a lower, more rounded design. The result pleased most of the group.

Murray also realized—perhaps as a result of the rocket-ship-like drawings that came out of Hinsdale—that he needed someone onsite who could draw concepts. As it turned out, Mel Van Buskirk was a talented graphic artist.

"Before he knew it," Murray wrote, "Mel had an almost full-time job making cutaway drawings and artist's renderings of our machines. . . . After three or four years, he let it be known that he really didn't like that work and wanted to get back into development. I felt, right or wrong, he . . . was most valuable in his current work. He resolved the situation by retiring."

That Axial-Flow Look

▶ In 1971, a new design was created for the Axial-Flow combine. The early versions looked much like the 15 Series. The final design was drawn by Harvester industrial designer Jerry McKirk, with Harvester industrial designer Gregg Montgomery coming in on the tail end of the project. Montgomery said the combine designs prior to the Axial-Flow lacked continuity and thought. "The IH Axial-Flow, on the other hand, was a very well-integrated design," he stated. "[It] looked as though someone had given great thought to how all the pieces worked together and fit together." *Gregg Montgomery Collection*

THE SURPRISE

In July 1972, Murray was told that one of his people discovered an unusual New Holland combine parked at a truck shop at Geneseo, Illinois—only about 18 miles east of East Moline.

Murray hurried over to look it over. He took photographs and did everything short of disassembling the machine. The combine was definitely a rotary combine, with a design that closely resembled the Harvester experimentals. The key difference was the New Holland machine had twin rotors mounted side by side.

"As an additional humiliation," Murray wrote, "it was on a big IH truck-trailer rig, the kind our management would never let us have."

The machine sat in the parking lot for the entire weekend, and the Harvester group took full advantage of it. "No one ever came out and accosted us," Murray wrote. "Either they just didn't care, or it was deliberate bringing the machine through the Quad Cities and parking it in Geneseo."

Murray wrote that this was the first time he had any inkling New Holland was working on a rotary combine. "We had been watching Deere and Massey with their 'secret' development activities but had completely ignored New Holland," he recalled. Considering the fact that Rowland-Hill had gone to work for New Holland, Murray admitted that, in hindsight, he should have been watched more closely. "There were a few rather embarrassing questions asked by our engineering management," he wrote.

"Very quickly we put together the fact that when Bill Rowland-Hill left IH East Moline, he had apparently gone directly to New Holland with a proposal for a rotary combine. Immediately our people began to put together a picture of what had happened. Bill had dug into it with a vigor, even applying for a patent with DePauw and Karlsson. Our people remembered that he tore the test machine apart with a determined purpose they couldn't understand and that prints of layout drawings were made and disappeared without explanation. We gradually put all this together and presented it to the legal people, realizing that the state of the machine we had seen indicated possible production before we were going to be in production."

Harvester's lawyers filed suit against New Holland's owners, Sperry-Rand, in 1973.

During all the legal action, Harvester combine engineers Bob Francis and Tom Stamp found two farmers running an experimental New Holland twin-rotor in Imperial Valley. Murray wrote what Francis had to say:

"It was just sitting out there so we went out to look it over. We found it would run so we cut a half tank or so of grain. Then I thought we should see what was inside so we went to work and soon had a couple of concaves out on the ground. Then, here came two farmers wanting to know who we were. Tom was tempted to just leave and say nothing but I said, no, we will have to 'fess up.' They were quite upset, we left our names, tried to visit a little, and left."

The farmers went straight to the New Holland engineers and they called back to the office. The New Holland office folks called their legal teams and the whole matter was brought to Murray's attention in a few hours. Bob had called Murray to alert him to the situation. New Holland threatened to sue, but it all calmed down.

"IH management," Murray wrote, "was not too happy with 'that reckless bunch over there at East Moline.'"

The court case tied up some of the engineering team's time. Murray recalls that Dick DePauw, in particular, had to put some time into the matter.

Behind the scenes, a patent war was quietly fought.

The original Axial-Flow combine patent was filed on August 30, 1966, and is assigned to Edward William Rowland-Hill of New Holland, Pennsylvania, and Richard A. DePauw and Elof K. Karlsson of International Harvester.

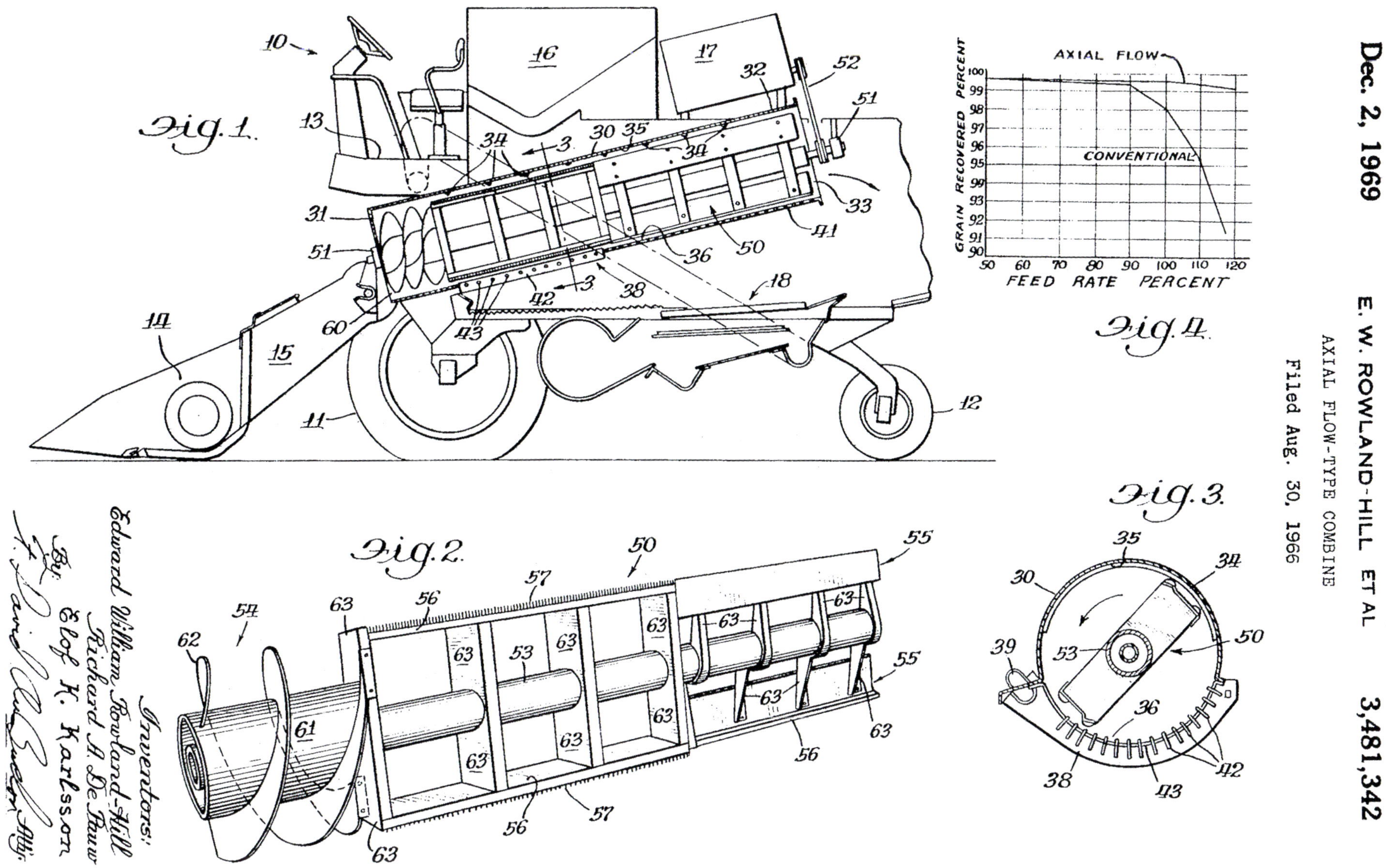

The First Axial-Flow Patent

▲ This is one of the first patents for the Axial-Flow combine as a unit. Filed in December 1966, it bears the names of Elof Karlsson, Dick DePauw, and Edward William Rowland-Hill. The patent was issued to International Harvester, but Rowland-Hill had left the company in 1965. He would later file a patent for a twin-rotor design that New Holland released in the mid-1970s. The wrangling over who owned the patent rights to the Axial-Flow continued into the 1980s. The end result was sealed by the courts. Today the two companies reside under one corporate entity. *U.S. Patent Office*

This was filed *after* Rowland-Hill left Harvester and doesn't make any reference to New Holland or Sperry-Rand.

After Edward Rowland-Hill left he filed a steady stream of patents relating to the rotary combine. A flurry of them were filed on November 24, 1969, and some were listed as a continuation of patent 790,145 filed on January 9, 1969, titled, "An Axial Flow Threshing and Separation Machine." These were granted in 1972.

Rowland-Hill filed more patents in the early 1970s. He filed patent 3927678 on May 2, 1974. The patent is for concaves used in axial flow threshing, which directed crop material through the machine.

On June 3, 1974, Rowland-Hill filed patent 3916912 A, covering a rotary combine with a rear-discharge beater. The patent refers mainly to how the airflow through the beater assembly is controlled. Interestingly, the patent does not reference the original Axial-Flow patent filed in 1966.

After less than two years of legal wrangling, the matter was settled out of court on January 22, 1975. The court records were sealed because trade secrets were revealed during the proceedings.

New Holland introduced the TR70 twin-rotor rotary combine in June 1975.

Murray said he sat next to Rowland-Hill at the annual ASAE dinner in 1967. "I found him to be a very engaging man, not at all the ogre we had projected," he said

"In all due respect to the integrity of Rowland-Hill," Murray also wrote, "all our suspicions were circumstantial, none were ever proved."

Harvester did not take the situation lightly. In preparation for the press release about its new Axial-Flow combine, the company provided a canned answer for people to use when questions arose about the lawsuit.

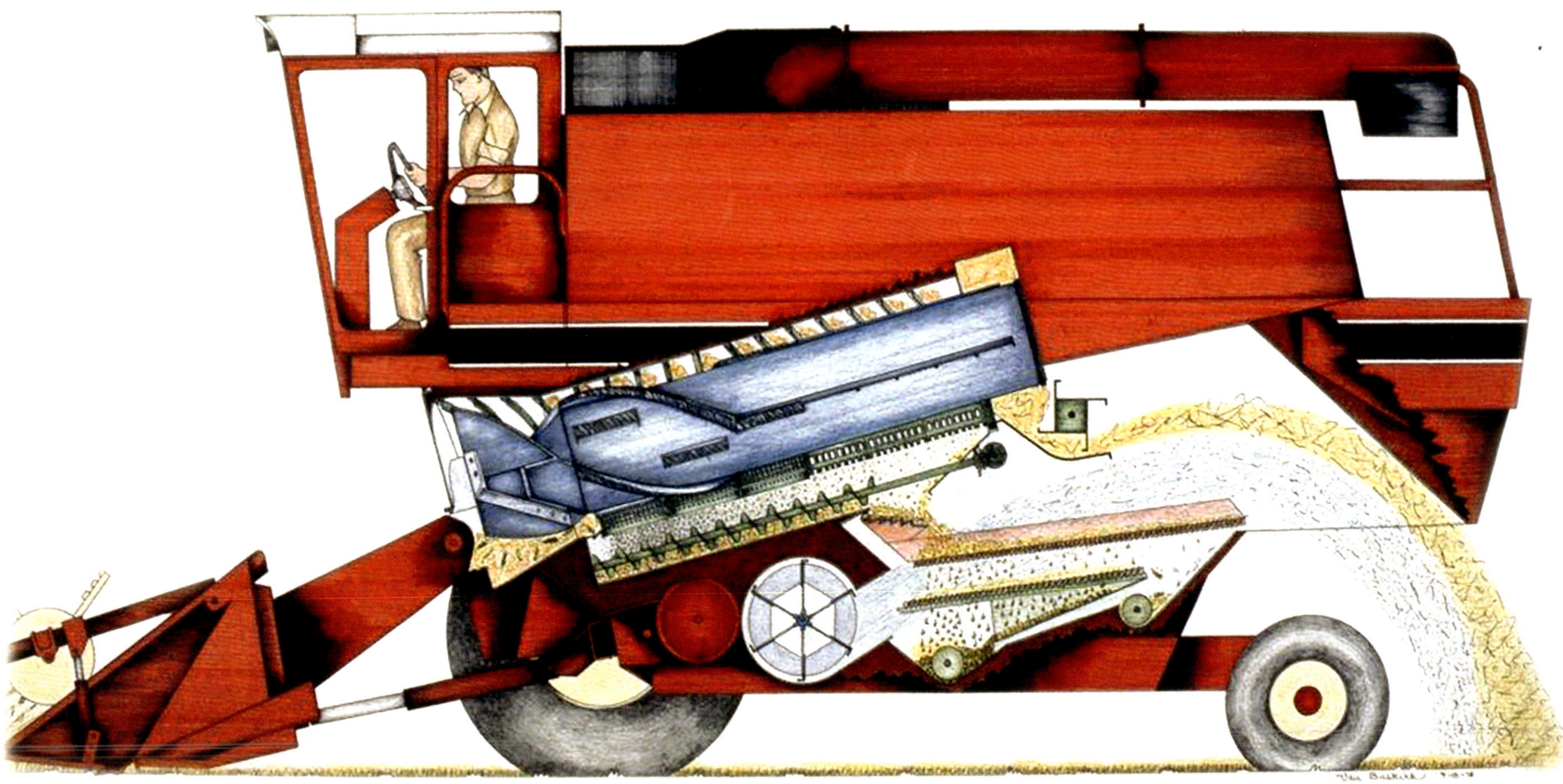

Early Cutaway

▲ This drawing by Mel Van Buskirk shows the rotor placement in the International Harvester Axial-Flow design. *U.S. Patent Office*

"The question relates to a lawsuit between International Harvester and Sperry-New Holland. Yes, there was a lawsuit initiated by IH against Sperry-New Holland in 1973. It was settled out of court in 1975 by mutual agreement. While we are not at liberty to discuss details of the suit or the settlement, we can say there resulted absolutely no encumbrance to our design development. We have patents on key elements of our design, and the combines we are announcing to you today are precisely the machines we wanted them to be. We invite you to compare them closely with any other combines on the market."

If further questions were asked, the Harvester people were instructed to respond, "No comment."

The story has an interesting footnote. Deere & Company filed a lawsuit against Morgan M. Finley, the clerk of the Circuit Court of Cook County, Illinois, asking for access to the sealed files from the 1975 case between Harvester and Sperry-New Holland. On December 30, 1981, the court ruled that Deere would be allowed to see the general court files, but not any of the material deemed confidential.

The technology developed by New Holland and Harvester in the early 1970s would be of interest to Deere into the 1980s.

THE WAR HORSE

Elof Karlsson retired before Murray came to East Moline. Retirees typically stopped in periodically and chatted a bit about old times for a few minutes. And then they would leave.

Not Karlsson.

He would come in and carefully analyze the machines in the shop. When they didn't meet his standards—which was pretty much all of the time—he would visit Murray in his office and explain seriously, in his thick Swedish accent, where and why the team went wrong. "Sometimes he became quite dramatic," Murray wrote, "absolutely adamant that we were going in the wrong direction and would ultimately jeopardize his retirement."

Murray eventually had to tell Karlsson that while he understood his intentions were good, the results were not helpful. From this point forward, Karlsson was welcome to visit—but not to study the machines.

"Elof was a bit offended," Murray wrote. "Sometimes it's awfully hard to get the old fire horse to stop following the fire engine."

"BAMFOOZLED" BY THE FRENCH: 1972

Testing a combine overseas is a great way to not only experience conditions in other countries, but also buy additional testing time. Shipping a machine to Australia gained an extra season of harvest for testing, and conditions were comparable to North America. In summer 1972, Harvester brass decided to send a modified CX-18A, dubbed CX-18AF, to France.

International Harvester's French division was building its own combines and wasn't tremendously interested in what its North American counterparts had to offer. Murray's boss told him that he should "get over there and learn what that situation was like." Murray did so grudgingly, and wrote that he wasn't terribly thrilled to be going in place of the test engineer, which meant he would be responsible for running the machine.

Murray hadn't traveled overseas since 1945, so the trip was a grand event. He was given two goals with the CX-18AF: build U.S.-French relations and, above all, keep the CX-18AF's presence a well-guarded secret.

Murray prepped for the trip with Midwestern diligence. He acquired standard-to-metric conversion charts, an English-to-French dictionary, a good map, Myra Waldo's *Travel and Motoring Guide to Europe*, and some notebooks. He also thought anti-diarrhea medication would be good, but his boss assured him that if he adjusted his diet with the right combination of wine and cheese he would be fine.

1972 Axial-Flow CX-18

▶ In 1972, Harvester shipped an American CX-18 to France to compete head-to-head with the latest French offering. The results of the informal competition demonstrated both the Axial-Flow's potential and the need for additional development. This CX-18 is shown in Rio Vista, California, in 1972. *Case IH*

Jean Thieffry met him at the Paris airport and immediately took him "roaring" through Paris to the "seedy old warehouse" that headquartered IH France. After a lunch heavy on his boss' prescription—wine and cheese—Murray retired for the evening.

The next day, he was struck by the multinational make-up of the French IH engineering team. "They had men from all over Europe and from India and other unexpected places," Murray wrote. The men received per diem housing allowances, and many chose to live in conditions that Murray found shocking in order to pocket the extra cash. "Some were little more than converted chicken coops," Murray wrote.

Testing was done in north France. The fields were provided by one Mr. Perilliat, a Frenchman who owned a farm near the town of Marck. He had a large spread by European standards, but his shop equipment was in a shabby state that contrasted sharply with the neat, orderly operations familiar to Murray. Mr, Perilliat's wine, however, was good and generously poured. "He kept better track of his wine cellar than his shop," Murray wrote.

Mr. Perilliat's fields were much more impressive. The yields on the old soil rivaled any American farms Murray visited, and the crews proved hard-working and efficient.

Murray spent several days fine-tuning the American CX-18AF on several fields in the area. Some excitement ensued when the lightly muffled combine roared down the narrow French roads with the header attached. The roads had to be closed to allow the large, loud American machine to pass

The French 953

◀ The French entry in the 1972 tête-à-tête was a preproduction version of the 953, known as the FCX-32. The combine was a conventional straw walker design, quite similar to the American-built 915.

Jean Cointe Collection

through town. "I was sure I could see the shutters on the upper windows flapping," Murray wrote.

So much for building relations with the French—not to mention keeping the machine's presence under wraps. The combine had all the subtlety of a 300-pound offensive lineman in a bridal shop.

The French IH team was testing its FCX-32, a conventional combine later produced as the Model 953 and roughly comparable to the U.S. Model 915. Comparative field tests were done with the CX-18AF, and the IH France top brass arrived to observe. After a week of testing, Murray was frustrated with the results. "Our grain loss looked great but our straw throughput was not," Murray wrote. "I kept having this nagging feeling something was not right. I was being bamfoozled [*sic*] by the French."

Murray took a weekend to analyze the raw data and concluded that the CX-18AF was putting more grain in the tank with less grain loss (0.67 percent loss for the CX-18AF versus 3 percent for the FCX-32). The broken straw being expelled from the American machine was, however, an issue in Europe, where long stems were prized. The testing continued for several weeks and concluded with a grand test attended by most of the French brass and, of course, their engineering team. The machines were scrubbed and polished, and a New Holland/Clayson 1550 and a Claas brand also were brought in to participate. The goals and features of each IH machine were presented. Then, everyone would be allowed to operate the machines.

Lunch was an equally grand affair, "with wine always high on the agenda."

For both teams, the test proved to be limited by their own high opinions of their machine. The French concluded their machine was better, and Murray concluded the CX-18AF was better.

The most eye-opening portion of the trip had nothing to do with the competitive interoffice relations of the time. Murray was appalled at the unfinished state of the CX-18AF. "There were far more mechanical problems than I could have imagined," he wrote.

The mechanical drives for the separator and the rotor were problematic, requiring daily cleaning and alignment. The unloader tube was impossible for one man to swing into the unloading position. Bolt-head sizes appeared random.

Murray noted 116 issues he considered "quite serious."

"Our adventure did likely have one affect," Murray wrote. "It made us face up to our own shortcomings."

The theme continued throughout the remainder of 1972. The combines were tested in the field and in the garage. The results were consistent: the Axial-Flows harvested more efficiently than anything on the market . . . until a mechanical failure shut it down.

Unless these challenges were overcome, there would be no more need for secrecy.

DARK SECRETS AND DEEP POCKETS: 1973

Harvester had three different experimental Axial-Flow combines running for the 1973 season: the CX-14, CX-18, and CX-18A. The 14 had the 30-inch rotor, while the two 18s had the 24-inch variety.

The test engineers worked these machines mercilessly, testing hard both in the garage and on crops in Illinois, California, North Dakota, and Iowa. Security on the machines remained as tight as ever—the specific locations of the test sites were heavily guarded details.

"We couldn't tell our families or our wives or anything where we were going," Dave Gustafson said. "We literally could not tell. The chief test engineer didn't really know where the machines were."

Which isn't to say company policies were able to trump family loyalty.

"I told my wife," Gustafson added. "We all told our wives."

The company had a way to deal with *that*, as well.

For key tests, product engineers and mechanics were flown first to a neutral site. Someone met them at the airport and handed them a phone number and another plane ticket.

They would get on another plane and would then end up at the test site.

Leaks were inevitable once the machines were out being tested. Longtime dealer Gerald Heim recalled his first encounter with an Axial-Flow combine.

"We got wind of an Axial-Flow or some kind of a combine being tested, and I don't remember what year exactly, but it was a year or two before they were introduced," Heim said. "We found a guy by the name of Rip Van Winkle that was running a new style combine. It had 915 on it, so it looked like a regular combine, and we found that probably 75 miles from home. We went down there and tried to see it, inside of it, and they had it all padlocked up, and it was quite a little problem convincing those people to let us look in it, but we finally got to see it."

Jim Minihan joined the company in 1970, working for a retail location in Austin, Minnesota. He quickly worked his way up, first as a service manager based in Mason City, Iowa, and then as a service manager in Chicago beginning in June 1973. He first heard of the machine while working out near Plainfield, Illinois. His boss told him he had something he should see.

"We ended a few miles north of Plainfield, Illinois, and off of Highway 59, doing soybeans," Minihan said.

His boss told him, "This is coming into the future."

"I was amazed at what it was doing at that time," Minihan said. "That was 1973. That's the first time I saw the rotary."

The first test of 1973 took place in February near Green Park, Illinois, where a CX-18A was compared to a model 915

Field Test Engineers

▲ The Harvester test-engineering team, out in the field near El Centro, California, in May 1972. From left to right: Gene Donahue, Jack Baraks, Dave Gustafson, Dan Seits, Augie Desplinter, and Tom Moore.

Dave Gustafson/Case IH

1973 Benchmarking

▶ In 1973, the Harvester test-engineering group benchmarked the performance of the CX-18 in Sikeston, Missouri. *Dave Gustafson/Case IH*

and a 503. The performance of the Axial-Flow was 35 to 50 percent better than the other Harvester models, but the report stated grimly, "mechanically the machine was horrible."

Testing continued, with engineering taking three steps back to gain one forward.

A key point occurred when the group had the Hinsdale shops build a full-scale mockup of the operator's area. The engineers, product committee, and general managers all could physically inspect the planned cab and the sightlines to both a grain and corn header. With this vantage point, the team determined that the windshield needed to go down farther. The "deep pocket" design that resulted greatly improved visibility.

The Competition

▲ During the benchmarking at Sikeston in 1973, the CX-18 was compared to a John Deere 7700. The output of each machine was carefully measured. *Dave Gustafson/Case IH*

Catching the Output

◀ Bags caught the material coming out of the back of the machines and the cleaning system. Harvester technician Bob Griffin is in the vest; engineer Dave Gustafson is in the maroon jacket and yellow cap. *Dave Gustafson/Case IH*

Test Caravan

◀ The yellow trailer holds all the tools for the test group, and the red machine is a cleaning system from a Model 80 combine that was used to separate out the loss. This team logged hundreds of checks to benchmark performance of the Axial-Flow. The results encouraged the company that its machine would have an edge over the competition. *Dave Gustafson/Case IH*

TORQUE-SENSING DRIVE

One challenge with the early Axial-Flow machines was the development of the drive system for both the wheels and the rotor, which proved problematic well into the project. Variable-speed belts were used to drive both the final drive and the rotor. In production, hydrostatic ground drive became standard.

The technology was a much more sophisticated version of the centrifugal drive system used on snowmobile clutches. This was the first time a system like this was implemented in a combine, and development was a major challenge. As Gustafson recalled, development of this drive system started in 1970 and was spearheaded by engineer Gary Drayer. Gustafson worked closely with Drayer, providing field and lab data and building test stands.

The Harvester engineering team developed its own transducers to monitor output and worked closely with its vendors to fine-tune belt materials and tension as well as cam materials and angles to optimize performance and reliability.

In the end, the system proved an effective solution that lasted a long time.

"Tens of thousands of hours of test stand time have been used to first develop the drive system and then to upgrade as machine horsepower has increased," Gustafson wrote. He noted the fact that original 1440 had 135 horsepower and later models equipped with the drive had more than 400 horsepower. "This high-performance drive was a very big part of the Axial-Flow combine's success."

"Gary Drayer (deceased) was the first designer on this drive, a fine fellow, who was one of the many people dedicated to the Axial-Flow development process. I like to think that it was his ideas that placed all the drives on this machine in the manner in which they were placed. Quite a challenge because the main drives were all new and unlike any on a conventional combine. As a Test Engineer I worked with providing him the field and lab data he needed for the design and then building the early test stands to validate the work."

Belt Wizardry

◀ One key development was getting the final drive and the rotor drive functioning properly. This image shows the torque-sensing drive set up to record data in early development. In this picture it is being used as a ground drive in addition to the rotor drive. The ground drive application was later dropped and released in production as a hydrostatic drive. *Dave Gustafson/Case IH*

CX14 on Test Stand

◀ This 1971 experimental combine is undergoing drive-endurance testing. Dynamometers were used to load the rotor torque sensing and all the other drive components. A Harvester V392 gas engine was used to subject the rotor drive to the pulses of the engine. The engine ran under such heavy loads that its exhaust manifolds turned cherry red. Once the Axial-Flow program was approved, testing went on 24 hours a day, 7 days a week. "I can't even speculate how many gallons of gasoline ran through this engine," Dave Gustafson wrote. *Dave Gustafson/Case IH*

Rotor Testing Stand

▲ The rotor torque-sensing drive was tested on this stand. A combine diesel engine was used to stress the rotor drive. Tens of thousands of hours of testing have been done on stands like this. *Dave Gustafson/Case IH*

Data Center for Rotor Testing

▲ Data on the rotor test was gathered in this computer center, which was completely automated and had a flight recorder that was able to recall all events that occurred at a time of failure in any of the components. By 1973, Harvester was pouring all the resources available into developing the Axial-Flow combine into a high-quality production machine. *Dave Gustafson/Case IH*

LOGICAL AND SIMPLE: 1974

Gary Wells would become part of the Axial-Flow team in 1974. Sharp as a tack and eager to make a mark, Wells had spent most of his career working at East Moline, doing testing and overseeing quality and reliability issues. In 1972, he had moved up and worked in Hinsdale for one year, where he distinguished himself and moved up to the corporate office in Chicago.

Colleagues of his liked to say he was often the smartest guy in the room, and was happy to make you aware of that fact. The first time Wells called engineer Don Watt into his office, he offered a memorable line as Watt left. "Young man," Wells said to Don, "Product development is far too important to be left to engineers."

Wells described April 1974 as a difficult time for the IH combine line.

"We were experiencing a perfect storm in our combine business," Wells wrote in his memoir. "Market share was only in the high single digits and dropping, and profits were nil. Moreover, confidential third-party surveys that we had commissioned, involving telephone interviews of hundreds of farmers, indicated the brand preference of future combine purchasers would favor us in fewer than 10 percent of instances."

The future was bleak for the existing line of IH combines. John Deere's Model 7700 machine had been introduced in 1970 was performing well in the market, and management expected the New Holland rotary combine to be on the market in 1975 or so.

"Since we were at the point that almost any success we could wring out of a big new gambit to accelerate development of the rotary concept would be better than the dire outlook we had, the decision was taken to go for it," Wells wrote.

In 1974, the key objective for the combine group was to get the machine ready for market. As part of that push from the head office, Wells was appointed to the new position of project manager of new combine development. His job was to draw a path from the Axial-Flow as it existed in April 1974 to a machine that could be sold to the public.

That chart was known as a "critical path plan," and it was the first such plan that had been applied to the combine engineering group. Don Murray recalled it as a helpful tool.

"Gary developed a critical path chart that covered one whole conference room well," Murray wrote. "It did a lot to alert us where our bottlenecks might be."

The entire combine engineering team was dedicated to the Axial-Flow—no more development was done on the current line. The prototype Axial-Flow combines were sent out into the field for as much testing as possible. For the 1974 season, five CX-18 and CX-18A experimentals and at least one CX-14 were tested in the field in as many different crops as time allowed.

In July 1974, the name "multi-pass" was dropped in favor of "Axial-Flow." The engineers were disappointed, and even Murray had no clear knowledge of why or even when the term was dropped. He speculated the sales group didn't like the term. "In July the term disappeared, replaced by 'axial flow,' and poor old 'MULTI-PASS' was never seen again," Murray wrote.

The name debate aside, the focus for the project was on production, which was targeted to begin in limited fashion in 1976. "In that draft," Murray wrote, "we were becoming quite realistic."

On November 14, 1974, Harvester vice president Omer G. Voss addressed the Sacramento Rotary Club to talk about Harvester agricultural's role on an international stage. His speech promoted Harvester as one of the world's premier manufacturers of agricultural equipment as well as trucks and gas turbine engines. It also provides some of the expected compliments to California and its role in history.

Caroll "Red" Gochanour

▶ One of the stalwart test engineers for Harvester is shown here on a Farmall 460 tractor. Gochanour was hired in the late 1940s and a few years later became one of the first Harvester test engineers. He logged thousands of hours testing Harvester combines—many of those hours with the early rotary models. *Gochanour Family Collection*

"If agriculture elsewhere in the world could rise only a fraction of the height already achieved here," he said, "there would be no scarcity of food anywhere."

Voss pointed to the increasing global interdependence of markets and the desperate need for "continuing devotion to research and engineering aimed at steadily higher levels of farm productivity."

Harvester's Axial-Flow, in the meantime, was getting closer to becoming the machine that could meet the lofty goals spread by the likes of Voss.

Carroll "Red" Gochanour had been testing Harvester equipment since the 1940s and spent his early days riding cross-country in an IH Travelall to test sites. "Anyone who couldn't play cards had to drive," he said. Gochanour despised airplanes, so he drove to the test sites.

Gochanour recalled testing the Axial-Flow in fall of 1974. His crew and Camiel Beert were testing a new low-profile 915 in San Antonio, Texas, in August and September. A CX-18 was brought down to harvest corn.

Gochanour was struck by the loss with the 915. No matter how they set it up, the ears that went through had seven or eight kernels left in the shucks that had surrounded the ears.

The CX-18 left the ears in pristine condition—and almost perfectly cleaned.

"The loss was half or less than what it was with the 915," Gochanour said.

By winter 1974, the increased manpower and budget had made dramatic improvements to the Axial-Flow, and the basic mechanical parts layout and position was set.

"It all looked so logical and simple," Murray wrote, "we wondered why it had taken so long to arrive."

Test Comparison

▶ Development continued in 1974, when this image of the early experimental model tested against a couple of late-model 15 Series machines.

Dave Gustafson/Case IH

The Vision

◀ In 1974, IH placed a lot of resources toward on finishing the Axial-Flow. With more people and more dollars, the team could envision the machine it would create.

Wisconsin Historical Society #114758

A NEW NAME: 1975

Testing for 1975 continued, and the field performance dramatically improved. Eight test machines harvested more than 16,000 acres. By the end of the year, the team concentrated on endurance testing rather than raw development.

Reliability remained an issue. The engineering group started tracking that in 1973 by dividing operating hours by operating hours plus repair hours. It measured three model 915s and found they averaged about 95 percent reliability.

The seven prototype Axial-Flow combines in the field in 1975 had an average reliability of 75 percent. For a prototype machine, this wasn't bad. The focus for the next year would be improving that.

Reliability was a major concern. The machines had terrible issues with that in the early 1970s, and many nagging problems remained unsolved as the production neared. Test engineer Ken Johnson brought this up repeatedly.

"I remember that Ken as test project engineer repeatedly expressed our concern about these reliability performance concerns. Don Murray assigned Product Engineer Don Olmsted specifically to the resolution of machine reliability," Dave Gustafson said.

"With Murray's support, Olmsted's management skill, design team efforts, and round-the-clock test stand operation, the machine was introduced with outstanding reliability. There were reports of custom cutters operating these early machines a harvest season with no repairs."

The names of the machine—after much debate and research—were also decided upon. Boeing's model was something that was looked upon favorably, with progressively larger models having two progressively larger final digits (707, 727, 747, etc.). Frank Farleigh in the end made the call that the names would be 1440 and 1460. The CX-18A became the 1440, and the CX-18 the 1460.

In October 1975, Harvester CEO Brooks McCormick signed the order to authorize production of 300 Axial-Flow combines: 100 1440s (formerly CX-18A) and 200 1460s (CX-18).

After 20 years of development, Karlsson and Van Buskirk's dream was coming of age.

Australian CX18A Testing

▶ Combines have to be tested in the real world—conditions change too much for any garage laboratory to adequately simulate real operation. This means combines and test engineers travel the world, following the harvest. This CX-18A was being tested in Australia.

S. Tomac Collection

Test Engineers

▲ In 1975, the Axial-Flow experimental machines were tested in Australia from January to June, shipped to California, and then shipped east and tested in Hamlin, Texas, and Thomas, Oklahoma. One of the machines went to Startford, California, for testing, and then was shipped back to East Moline and then to Phoenix, Arizona, for a rebuild and more tests. In Tucson, Arizona, in August, it suffered "many mechanical problems." August was a busy month, with machines tested in Brawley, California; Gurley, Nebraska; and Voss, North Dakota. After more testing in Brawley; San Antonio, Texas; and Burlington, Iowa, in October, the crew was back in the garage burning the midnight oil to develop the rotor. They closed out the year testing in Eugene, Oregon, and Sikeston, Missouri. Test engineers didn't get much rest. The crew above is, from left to right: Tom Moore, a local Australian farmer, Dave Gustafson, J. Francis, and Ken Johnson. They are in Darlington Point, Australia, on May 19, 1974.

S. Tomac Collection

WHO BUILDS IT? 1976

Once the decision was made to release the Axial-Flow combines, the thorny question of building them arose. Harvester built experimental and preproduction machines in a variety of places. In the early 1970s, East Moline engineers built many of their experimental combines right on site, as they had an extensive shop at their disposal. The East Moline prototype shop was staffed with about 100 people—machinists, welders, sheet-metal benders, and even a blacksmith.

The shop also had a photography studio and was stocked with parts manuals.

Fort Wayne Experimentals

▶ In 1976, the engineering team turned to the Motor Truck Engineering facility in Fort Wayne, Indiana, to build some of the early prototype machines. The results were reportedly solid. This is one of those early machines at work.

Dave Gustafson/Case IH

Fort Wayne Experimental CX-14

▶ Note that this prototype bears the 1440 moniker. The graphics are in place and the machine looks ready to go to production.

Dave Gustafson/Case IH

Many of the prototype Axial-Flow combines were built at this in-house facility at East Moline. Some were built at a similar facility in Hinsdale.

Traditionally, once a machine reached a more advanced stage of testing and required multiple units, a run of five or more machines was built by a manufacturing plant with extensive input from the engineering team. As one might imagine, manufacturing plants already under siege to meet production quotas were not keen to hand-build the engineer's latest creation.

Throughout the 1970s, manufacturing was putting up more and more resistance to building late prototype and preproduction machines. East Moline plant manager Curt Harter had become somewhat of a burr in the side of the combine team. When the time came to build the 1976 prototypes, he suggested to Murray that the East Moline shop, rather than his plant, build the prototypes.

Murray solved the issue when he found that the Motor Truck Engineering facility in Fort Wayne, Indiana, had room to assemble five combines. His group supplied engineering drawings and built many of the small detail parts, and the Fort Wayne truck group built the rest. The result was five combines that Murray described as "well-built." These machines were considered preproduction models. Murray said they were not sold, but were used for primarily for marketing.

For the 1976 season, a total of 17 machines were in the field, 12 of which were built by the engineering shop at East Moline; the other 5 were built in Fort Wayne. The model list is interesting. "The result was in 1976 we had 17 machines in the field," Murray wrote, "including two CX14, two CX-21, the new small combine, and one CX-38, the pull combine in the CX-14 class."

Murray believed that having the Fort Wayne truck group build prototypes was not good for the East Moline manufacturing plant, as they missed an opportunity to learn how to build the new machine.

In hindsight, he was right.

CX-21 Experimental

◀ This American-built CX-21 is shown in Australia bearing the Australian design comb.

S. Tomac Collection

CX-38 Experimental

▶ CX-38 was the code name for the experimental version of the 1482 pull-behind combine. Production was authorized in 1976. Here, two of the machines are shown at work in 1978.

Dave Gustafson/Case IH

THE EAST MOLINE PLANT 1920s–1980

By Lee Klancher

International Harvester opened negotiations with Moline Plow Company to purchase six acres of land in East Moline, Illinois, in 1926. Word quickly spread through the small town, igniting a firestorm of speculation. The *Moline Dispatch* reported the deal was under negotiation and estimated as many as 1,000 men would be employed at the new facility.

By year's end, Harvester had purchased the six acres as well as another 83-acre parcel. The news was so stupendous that Cyrus McCormick himself released a statement about the acquisition. "The coming of IH to this community is regarded in many circles as one of the most important events that ever occurred to this community," McCormick said. He went on to describe their plans to build tractors, harvesters, threshers, and more on the property, located on the banks of the Mississippi and with rail lines running nearby. By 1927, McCormick told the local paper that within a few years, all manufacturers will be moving equipment on the mighty river.

A 450,000-square-foot warehouse was built that covered 10 acres. Completed on October 26, 1927, the facility included an area for assembling threshers, harvester-threshers, and corn pickers. In 1929, the warehouse was expanded, resulting in a building with than 1 million square feet of floor space under a single roof.

In 1933, the combine plant was built on the East Moline property. The building covered 90,000 square feet and cost $200,000. Production began in 1934, and in 1936 the plant was expanded with the addition of a steel storage building and a boiler room.

The plant continued to expand through the 1940s and more property was purchased in 1946. The piece of land bought at that time was originally owned by Deere & Co.

The rise of the self-propelled machines drove more growth at the East Moline Plant, which built a total 33,897 units in 1953. Another 47 acres were purchased in 1960, bringing the grounds up to 156 acres in total.

As other plants closed—notably McCormick Works—East Moline Works took on more and more production. That would be a recurring theme for the giant production facility. Old factories would close and their tooling would be torn down and rebuilt at East Moline Works.

In the early 1970s, the name of the facility changed from East Moline Works to the East Moline Plant. In 1972, the Harvester credit union moved to a new building. In 1973, a new blacktop parking lot large enough to hold 1,400 cars was constructed at the East Moline Plant.

Throughout the mid-1970s, the plant was steadily improved with a new concrete floor and an $11 million modernization program.

Brothers Steve and Jamie Horst worked for many years at the East Moline Plant. Both had various key roles in management, and both take great pride in their time there.

Steve and Jamie explained the attitude at the East Moline Plant.

"We always deliver," Jamie said. "It was always the plants over and above. If they set new goals, safety goals? Met them. Beat them. Manpower goals? Met them. Beat them. Every time corporate put initiative on the East Moline Plants . . . delivered. Done. Easy. Easy."

By 1977, that attitude would face one of Harvester's most difficult manufacturing challenges: building the all-new Axial-Flow combine line.

East Moline Plant 1944

▶ The East Moline plant had 1,000,000 square feet of covered space. By 1944, when this image was taken, it was running year-round building self-propelled combines, and the line was capable of building 24 machines per day. On March 17, 1944, the 10,500th No. 123-SP started down the assembly line. *Wisconsin Historical Society #7720*

WOMENS
REST ROOM
AND
LOCKER ROOM
FIRE
DO NOT
USE ON
ELECTRICAL
EQUIPMENT
3-8

BUILDING THE 300: THE LOONEY ROOM

Bud Horst started at the East Moline Plant in the 1940s. In the tradition of so many Harvester men, he worked his way up from the bottom. He wasn't college-educated, but he was fair and diplomatic and had a gift for connecting with anybody. He retired in the early 1970s as superintendent of the fabrication area, and as a well-respected member of the plant manager staff. He was also instrumental in building the dike in East Moline.

Once when Horst was superintendent, he was walking the floor and noticed that one of the operators, Nick, had a lot of trash near his machine. He called Nick over and asked him to clean it up. As Horst left, he heard Nick mutter a line in Dutch.

"Nick, I believe you had something to say?" Horst asked. "I didn't understand that."

"I don't think you really want to know what I said," Nick responded.

"No, I'm very interested. I'd like to know what you were saying."

"Well, I told you to go to hell."

"Well, I appreciate you telling me that," Horst said, "but I'd still like to get that trash cleaned up from your machine."

Horst's sons, Steve and Jamie, saw Nick in Arizona in 2013. According to Jamie, the encounter on the floor made a strong impression on Nick.

"I really respected your dad. There wasn't a better individual in that plant," Nick said.

Steve and Jamie would follow in Bud's footsteps at Harvester. Filling those shoes would not prove an easy task. "It was tough for us because you were compared to him all the way through," Steve said.

Steve Horst took a job as third-shift blueprint operator as a kid fresh out of high school. He interviewed in the daylight and went to work his first shift at 11 p.m. At night, the 2.4-million-square-foot East Moline Plant became a dark, looming cavern.

Steve searched for his station for hours. A guard stopped him and asked what in the hell was he doing? Steve told him. The guard escorted Steve to his workstation.

He arrived six hours late.

The next day, Steve's supervisor saw that nothing had been completed the night before. Union rules saved Steve from being canned on the spot.

"He thought I was a derelict and was going to ask me to leave but he had to give me my probationary time," Steve said. "[I] almost got fired on my first full night on the job."

Steve buckled down and was able to keep his job. He tried some college for a year, which didn't work out so well. He was drafted, served in Vietnam as a staff sergeant, and finished college after that on the GI Bill. During this entire time, he worked off and on at East Moline. By the time he graduated from college, his part-time work translated into quite a few years of seniority. The seniority meant a clerical position in the shipping department at East Moline that paid well and offered several weeks of vacation.

The East Moline plant manager, James Wegener, heard about Steve and his degree. He recruited him into management. Steve took to that well and worked his way up to the number-two spot in the East Moline Plant.

He built a reputation as a straight shooter who was unafraid to call it like he saw it. He didn't play politics. He was analytical and sharp and adhered to rules and order. People viewed him with a mix of respect . . . and fear.

His nickname was Crusty.

Steve's brother, Jamie, took a similar path to Harvester but had a completely different career. Jamie got by in high school. His big personality gave him talent for and a love of theater. He tried college and liked the theater and social aspects of that stage of life, but found that studying wasn't his bag.

Eventually Jamie took a spot at Harvester as an assembler and spot welder, which didn't suit him either. He pestered the

Experimental Build-Up

▲ Building of the first 300 Axial-Flow combines took place in East Moline and required massive changes because the entire plant was designed to construct straw walker combines. This image shows construction of a rotor. *S. Tomac Collection*

Preproduction Model

▶ By 1975, the preproduction machines were well finished.

Case IH

human resources department until they gave him a gig as a time study engineer.

Time study engineers determined standard times for jobs on the line.

"They were probably the most hated men in the plant," Jamie said.

Jamie's new boss didn't make matters easier. "He was a Napoleon wannabe," Jamie said. "When you walk in in the morning, he's been there for hours, dictating and looking at spreadsheets. There was never anything on his desk but a pack of Marlboros and a lighter."

He would check in on his people to make sure they didn't go to the bathroom during work hours. He spent a lot of time covering things in red ink.

He didn't cut Jamie any slack.

"You know, by the way, just because your old man was here, don't pull no shit with me."

"No, sir," Jamie responded.

"I'll ditch you just like anybody else if you're not doing something right."

The boss told Jamie that he saw people quitting early all the time. He wanted Jamie to cut the time rate by 5 percent. To do so, Jamie would have to confront an entire line of workers and find ways to make their quotas higher.

This was why people hated time study engineers.

Jamie pulled the general foreman and the superintendent together. He told them straight what he had to do. They walked the line. They analyzed and inspected. In the end, they found a few places to cut time.

They cut the rate by 2 percent.

Jamie went back to his boss, expecting to catch hell for making less than half the goal. The boss just chuckled, said he was surprised Jamie was able to cut that much.

Jamie passed that test and also learned his strength: working with people. He turned that into a career, working first as an industrial engineer and always as a guy who could build consensus and innovate. He also showed a gift for making the pursuit of quality a competition. He took busloads of people to ballgames and bars as a reward when their unit won quality contests. People loved it. His quality program became a model for Harvester, and he taught others how to do what he did.

Jamie and Steve also learned they could work together very effectively. Their styles were complementary. Jamie was the natural good cop. Steve was the enforcer.

"They knew that with the Horst name, you got what you paid for. They also knew that if I didn't get the answer that I was looking for, then I would turn them over to Crusty," Jamie said. "That was their worst nightmare."

By 1975, Steve was the production controller, running a team of about 100 people. Jamie was an industrial engineer, solving problems on the line.

Both would play key roles in building the first 300 Axial-Flow combines.

When the Axial-Flow combine came online, the East Moline Plant already had a full plate. They were building three different current-model combines. They were building grain and corn heads. They also built planters.

New machine builds were laid out in a separate warehouse. A team of industrial engineers, planning people, and process engineers would determine tooling and rigs and line layout. Parts would be specified and ordered. They often didn't fit and would be have to be reordered.

The guys at the East Moline Plant called the warehouse where the Axial-Flow was built the "Looney Room."

Manufacturing at East Moline had nicknames for a lot of things. Parts were nicknamed: the coffee grinder, the donkey dick, and the elephant trunk. Top brass in Harvester weren't terribly fond of these names. The East Moline crew was told to stop using the term Looney Room.

They quit using the name in reports and in certain meetings. On the floor, the name thrived.

"We called it the Looney Room because we were trying to build and put together an Axial-Flow combine," Jamie said. "We were trying to put this thing together so we get an idea of the time that it was going to take to build it down the assembly line. It was hectic. It was chaotic, hence the term Looney Room that we were out there and part of for months, just squirreled away in this area."

The Looney Room earned its name with the Axial-Flow combine. The rotary system required all-new manufacturing processes. The people at the plant weren't even entirely sure the Axial-Flow existed prior to laying eyes on one.

"We heard that, 'Oh, there's a new rotor coming,'" Jamie said. "Straw walkers are going to go away. That's all we knew until this thing rolls in and we see it firsthand."

They had roughly a year to assemble an all-new production line and produce 300 preproduction units of an all-new combine.

To make matters worse, this would be a one-time build. The group would have to design a line to build a machine, build a couple of test examples, and then set up the factory to build 300. Once that was done, the whole thing would be torn down and packed away. When—and if—the decision was made to go into full production, the line would come back out of mothballs.

Steve Tyler recalls that the work required to put those 300 into production was overwhelming.

"Then there was a massive facility change that had to be undertaken," Tyler said. "We put in a new combine line—number five combine line. Built a new wash and paint system for the combine involving some overhead man carriers and underground robotics for the painting. It was a big undertaking, from a facility and a manufacturing ordeal."

"It's just a crazy environment," Steve said. "You're trying to do stuff and put stuff and notice the clock is ticking."

Building the rotor proved to be one of the most difficult challenges. Straw walkers were quite forgiving when it came to tolerances and simple to manufacture. Not to mention the fact that the East Moline plant had been building straw walkers since the 1940s. They were easy to build.

"You put a bunch of spots on a bunch of pieces of sheet metal and you line them in the fixture and you look them up and do a little crank and away you go," Steve said.

The rotor, in comparison, was all-new, and had to spin at a high rpm. The rotor had to be perfectly balanced or it would fail. This was new territory for manufacturing, and it was the heart and soul of the machine.

"Spin the rotor on a dynamic balancer," Tyler said. "Add weight by welding them. Get the rotor so that when it goes into the combine, it will spin and function at pretty high rpms, without causing the combine to shake and rattle and fall apart. It was a big deal for us at the time."

While development was going on night and day in the Looney Room, their staunchest competition was right across the street. Truthfully, they were closer than that.

The East Moline Plant backed right up to the John Deere combine plant. People could look out the window and see train loads of combines being shipped.

Keeping secrets in East Moline required dedication. Development could be done in confidence. The secret garage

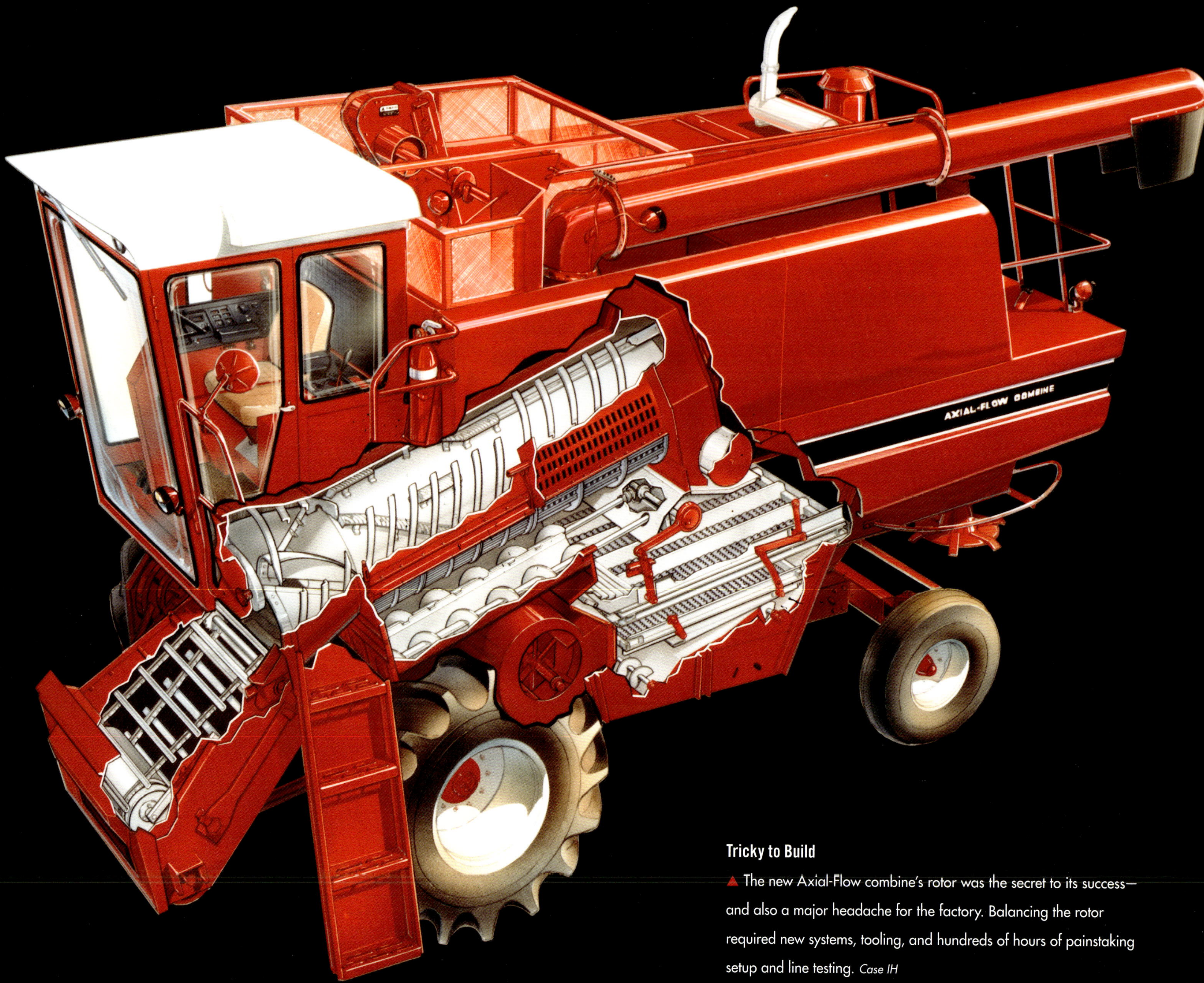

Tricky to Build

▲ The new Axial-Flow combine's rotor was the secret to its success—and also a major headache for the factory. Balancing the rotor required new systems, tooling, and hundreds of hours of painstaking setup and line testing. *Case IH* at the engineering facility succeeded in that, for the most part. Access was kept to just a few handpicked people in the early days. The news was kept to whispers and rumors.

The East Moline Plant employed 3,400 people in 1975. When you build a brand-new machine in front of that big of a group, secrets are impossible.

Management didn't even try to combat it.

"I don't think we were ever told that we couldn't talk about it, not at all," Steve said. "It certainly was a hot topic because it was kind of a make or break thing. If this thing doesn't work, what are we going to do?"

So the John Deere plant people knew Harvester was building a new machine, one that didn't use straw walkers.

They weren't impressed.

"Along with us doing all of this, you had the people right next door to us—John Deere, okay—laughing at us, a bunch of idiots that are going to bring out this rotor," Jamie said. "They can see why we have a loony bin because we're all going to be in it.

"The green guy is laughing at you all the time. We were just a sitting duck for them waiting for the avalanche to come crashing down on this little red combine."

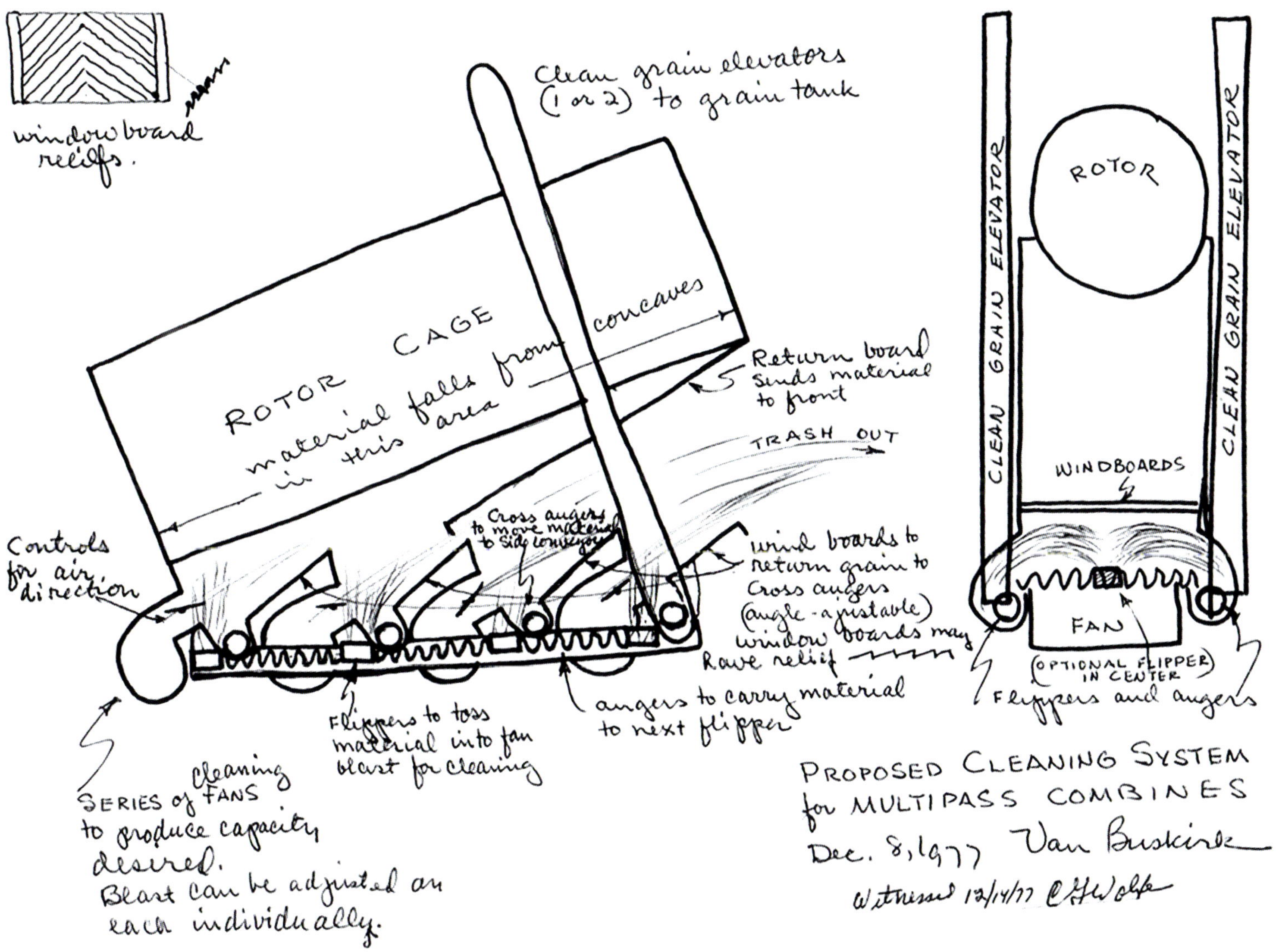

1977 Cleaning Unit Sketch

◀ Development of the Axial-Flow combine was a relentless process. This sketch was a proposal to modify the cleaning unit on the Axial-Flow with adjustable air blasts and windboard angles. Dated December 1977, the proposal is evidence that the engineering team continued to work to perfect the concept even after the first 300 units were out in the field.

Dave Gustafson/Case IH

On the line, people were concerned. If the new technology failed, their jobs were at stake.

Marv Wyffles was brought on to help get the combine line up and running. Steve said he helped people get over their fear.

Steve remembers Wyffles telling the plant workers, "You've got a chance. You have a once-in-a-lifetime chance to be part of something brand-new. This has never been done before. You can be like the first guy on the moon. You can build a part of the history."

Steve pointed out there were labor issues as well. The union watched all the time studies closely, making sure the times and rates met their expectations.

When the first Axial-Flow combines went out the door, they were tightly covered in canvas. Manuals were shipped separately to the dealers, not stored with the combines.

"Cloak and dagger," Steve said somberly.

Steve would drive right past the Deere harvester plant on his way to work. One morning, he could see legs sticking out from under a tire.

"I knew it was an engineer equipped with flashlights or clipboards or cameras under the tarp taking pictures of our new machine," Steve said.

One of the first examples sold went to Texas. A panicked call came into the East Moline Plant—the combine had caught fire.

Plant manager Matt Glogowski immediately chartered a plane to Texas. He also demanded engineering send someone with him. Dick DePauw, one of the key engineers on the project, went and took a quality-control engineer with him. Upon examination of the combine, they determined the problem was in the rear of the rotor. Manufacturing was using looser tolerances than engineering, and the space left around the rotor allowed straw and chaff to wrap around the shaft. The tangled straw would heat up and light on fire.

Back in the East Moline Plant, tolerances were tightened. The problem was solved. Work continued.

Preparing the production line happened in fits and starts. Parts were ordered and work was done, and then the regular production of the rest of the product line would interrupt things. Pieces were scattered to the winds.

When it came down to building the last few machines, the parts-ordering process left things a bit scarce.

"How many times do you think you could buy 300 toothpicks and six months after you bought them, account for all 300 toothpicks?" Steve said.

"You'd make them two or four times because parts wouldn't fit. You didn't make enough. You lost them."

In the end, the schedule was met and 300 preproduction Axial-Flow combines went out in spring 1977—on time. Which isn't to say all were perfect.

"Here you got this one-time build of 300 things," Steve said. "You think you want the last one?"

THE RED DEMON: 1977

The 300 1440 and 1460 combines produced in 1976 were sold to customers around the United States carefully picked by the marketing group. They chose customers thought to be loyal and would report back on the problems and successes. Most importantly, they were not to defect to a competitor's machine.

One customer proved to be a poor choice.

The green guys were doing all they could to get their hands on an axial, and it paid off in Minnesota. They convinced a farmer to part with his early production International Harvester Axial-Flow in exchange for a brand-new John Deere combine and a few thousand dollars in cash.

A few days later, Deere's chief test engineer called Don Murray and asked for a copy of the operator's manual and any other publications that went with the machine.

Murray played coy and agreed to send them over.

The John Deere man also asked if they could send over a test engineer to help set up the machine properly.

Murray declined.

He did, however, let the Deere man know when a field improvement package was released. But he wasn't a pushover—he made the Deere guys pick it up and install it.

Glenn Kahle was the head of the advanced engineering group at John Deere at the time. He was sent to East Moline to develop a Deere rotary separation combine.

"Deere was very conscious of the Axial-Flow development, and they wanted to make sure we weren't behind," Kahle said. The company established an advanced harvesting group for the task.

The International Harvester Axial-Flow combine purchased in Minnesota was used as part of a complete program designed by the Deere engineering team.

"For four years, we developed harvesting systems and one of them was based on the 1460 combine that IH built," Kahle said.

His team developed a complete program and not only tested that machine they bought in Minnesota, but used parts of it as a development mule. The most advanced machine

1977 International 1440 Axial-Flow Combine

▼ In 1977, after more than 20 years of development, 300 preproduction Axial-Flow combines were built. They were a mix of 1440s and 1460s. This image shows one of those 300.

Case IH

was a four-wheel-drive articulated combine that used a 1460 front end.

None of the Deere rotary combines would see production until 1999.

"We had a complete line of rotary combines ready to go," Kahle said. As he recalls, the investment required to produce them was about $120 million. In order for the machines to come close to the profit margin on the conventional Deere machines, the cost of each would have had to have been more than $100,000.

"The financial guys at Deere looked at their balance sheet and we were making tremendous profits on the [conventional combines] they had in production," Kahle said. "They had a good reputation, so they elected not to go ahead with it. Which was one of the smaller reasons I actually left and went to IH."

Kahle came to work for International Harvester in 1979 and played a pivotal role in the new research and development that would eventually create the Magnum tractor.

Murray reflected that from a dollars and "sense" perspective, the development of International Harvester Axial-Flow combine probably never would have made the cut.

"If we had known the real magnitude of our undertaking and the time and money it would have consumed, IH management would never had condoned it," he wrote. "The project would have died."

The Axial-Flow project succeeded because of the passion of the men in that secret shed who kept working doggedly to make the technology work.

"There was one other factor that was hidden in the wings," Murray wrote. "IH was continually in the shadow of the competition, primarily Deere, not entirely because of our combine designs but in the shadow nevertheless. We needed something spectacular, a breakthrough to get the farmer's attention."

The Axial-Flow took the combine group out of anyone's shadow.

While John Deere was playing catchup, the Harvester crew enjoyed shining a light on its new creation.

On June 2, 1977, the Harvester team hosted an event it called the "Quad Cities 300." All the regional sales managers and other marketing staff were invited to East Moline for two days of tours and familiarization with the new Axial-Flow combine. The engineers gave presentations.

Murray spoke about the development of the machine at the Society of Automotive Engineers Off-Highway Vehicle Meeting in Milwaukee on September 12 to 15. He and Dick DePauw, Jim Francis, and Ken Johnson presented the same information to the ASAE International Grain and Forage Harvesting conference in Ames, Iowa, a few weeks later.

The same month, the press was brought in for a grand presentation, as well. "Unfortunately it was drizzling rain during the field work but in true Axial-Flow form we harvested corn very satisfactorily anyway," Murray wrote.

The reaction to the Axial-Flow was phenomenal. Murray kept a log of 32 articles on the new machine. *Machine Design* magazine did a cover story about them. Farm Chemical Companies used photos of 1440s and 1460s in their advertising.

The wave of good press put the competition on edge. Deere and Massey produced technical rebuttals saying that they would have rotary separation if it was superior to their existing systems.

In reality, both were working feverishly to develop a machine that used rotary separation.

New Holland, for its part, touted its allegedly superior twin-rotor machine.

At the 1977 Farm Progress Show in Washington, Iowa, the competition had a chance to match up head-to-head. Camiel Beert, Harvester's most senior product man, was running the Axial-Flow. In fact, he was the company's first choice. His job was to travel all over the world, helping customers and dealers get the most of their machines. Beert understood the problems customers were having and would move heaven and earth to see their issues dealt with. He was legendary for calling engineers or management and telling them that a customer was expecting a call in a few minutes.

This didn't make him the most popular guy in the company. But the fact that he could do this strongly illustrates the combine group's customer focus at the time.

Beert was also a bit of skeptic early on. He had been running conventional combines since the 1940s.

"When we went rotary, I thought we were nuts," Beert said.

One of the first times he operated the new machine was at the 1977 Farm Progress Show. The conditions were awful. "Muddier than bloody hell," Beert said. "We just couldn't get very far because it was muddy."

Beert recalls asking Marv Wyffles, the assistant to the works manager, to go get him some rice tires so they could run in the mud.

"Are you crazy?" Wyffles responded. "I'm not your errand boy!"

After some discussion, the big tires were shipped to the show and mounted.

They worked perfectly. Despite the mud, the Axial-Flow harvested corn with flair. Beert quickly figured out how to set up the machine, and its performance was outstanding.

The John Deere got stuck.

At the time, the agricultural companies at the show used radios to communicate with drivers. One of the Harvester guys, Ralph, had figured out how to listen in on the Deere guys. At one point, Ralph gave the radio to Beert, saying, "You may want to listen to this."

"The Red Demon is climbing up on us!" said a Deere guy. "Go faster."

"It can't go faster," the operator responded. "I'll plug this thing!"

"Go faster!" the Deere guy said.

Test Engineer Camiel Beert

◀ Camiel Beert started with International Harvester in 1947 and eventually became a test engineer. He tested most of the combines the company built and traveled the world providing his input on IH and Case IH combines. He dealt with some of the thorniest problems and customers and became the go-to operator to get the best out of a machine. He's shown here testing a 503. *Case IH*

They plugged it.

After that, the Deere operators were instructed to stay away from the "Red Demon."

The Red Demon got the better of the Deere that day—and many more.

Regular production of the International Harvester Axial-Flow combine was authorized to begin in March 1978, and East Moline plant manager Matt Glowgowski decided the time had come to celebrate. "With all the staff wearing red blazers, Matt triumphantly drove the combine through a huge paper banner and all applauded," Murray wrote. "Morale was high throughout the entire factory."

The same could be said for the entire Harvester organization.

Elof's obstinate vision had survived long enough for the next generation to pick up the change. Don Murray's team made the Axial-Flow combine work so efficiently that the rest of the industry would be playing catchup for a long time.

The Red Demon had trumped the rest of the colors.

AXIAL-FLOW COMBINE

Chapter Five

1978–1985 THE AXIAL-FLOW DEBUTS

By Lee Klancher

"[The Axial-Flow combine] is probably the single piece of farm equipment that changed the face of agriculture more than any other."

—Don Watt, Case IH engineering manager, 1981–2014

Charles (Bud) Hoober starting selling IH equipment at his dealership in Intercourse, Pennsylvania in 1941. In 1965, his son, Charles Jr. (Charlie), became a partner in the business with his dad.

Tom Yohe is one of the family's long-time managers who became a partner in the business. When he came on in the 1970s, the farm economy was booming in no small part due to increased demand for exported grain, particularly to the Soviet Union. Sales were up for the Hoober business, particularly with tractors and implements.

The existing IH combines, however, didn't sell well.

"When I joined the company, we had a very small percentage of the combine business with the 15 Series," Tom said.

When the opportunity arose for Charlie and Tom to see an Axial-Flow, they jumped at it. "Charlie is a pilot and had access to a small plane at the time, and flew a couple customers to Ohio to see a preproduction model," Tom said.

The most popular combine in their area at the time was the Massey-Ferguson. "The customers were running Massey combines at the time and had been for a number of years. When they saw how the Axial-Flow worked, how simple it was, they said, 'We want one as soon as we can get one.'"

The Hoober dealership received one of the original 300 prototypes, a 1460—a machine Charlie and his family still have today. Rather than sell it, they used it to show their customers how it worked.

The demonstrations were a runaway success.

"We sold about everything we could get, at that point in time," Tom said.

"Yeah, and from then on, it just kept growing," Charlie added. "The thing that sold the combine was the grain quality and the [reduced] grain damage . . . and the field loss was less than our competition.

"We also had a very good corn head at that time that went along with this machine. That corn head came out with the 915s in 1975, and everybody liked that corn head because it

Axial-Flow Debuts

◀ The 1460 and 1440 were released for full production in 1978. The machine's performance was superlative. This image is from the 1978 buyer's guide.

Wisconsin Historical Society

Cold Harvest

▶ Pickings for farmers were terribly thin in the 1980s. Lower prices and historically high interest rates made 1980 the hardest year on the farm since the Great Depression. In this image, Dodge County, Wisconsin, farmer Larry Luhn, along with relatives and friends, picked corn in a field along Bancroft Road. *Wisconsin Historical Society #92822*

Petition for Help

▶ The tough times of the 1980s led many farmers to take political action. Jerry Zarecki (in hat), president of the Portage County Farmers Union, presents Wisconsin Congressman David R. Obey with a folder of petitions collected by Wisconsin farmers protesting the discriminatory dairy pricing provisions of the 1985 Farm Bill. *Wisconsin Historical Society #101245*

chopped up the stalks and it was a very aggressive corn head. The introduction of the rotor combine with that corn head was a very, very productive team."

The success of the Axial-Flow was tied closely to the growth of the Hoober dealership network. It carried Steiger tractors at that time—which was not yet owned by Case IH—and developed a good reputation with that tractor. The Axial-Flow came at the perfect time and allowed them to expand.

"We had Pennsylvania, New Jersey, Maryland, Delaware, Virginia, for the Steiger tractor," Charlie said. "When the Axial-Flow combine came out, we just picked right up and kept rolling, right with those same customers."

The early model Axial-Flows worked terrifically but like any brand-new technology, improvements were made in the first years of production. The machine needed better rock protection and a faster way to carry the grain away from the threshing mechanism.

"There's just always one part of the machine that doesn't quite keep up with the rest," Charlie explained. "It's the elevators and then the grain bin and the horsepower, and then the sieves have to get bigger, then the bushings in the sieves. The back axle's not heavy enough, so you've got to beef that up. It just always pushes the problem to another area. That's just the way it is.

"The key to being successful with an innovative product is following the market very closely and responding to the market's problems."

Charlie continued: "Everybody thinks the engineers design and develop, but really the farmers design and the engineers develop and perfect a combine. The time that was spent in meetings asking farmers what they like to see in a machine is very, very important."

At the close of 1978, Harvester sent out a mail survey to every Axial-Flow customer to learn how the machine performed. Satisfaction was very high, but Harvester also learned about some areas for improvement. Harvester made the improvements while also making certain they could be retrofitted to models already out in the field.

"You can take an older machine and update it to the new machine," Charlie explained, "so what happened was the resale value of the older machines was very, very well maintained, better than the competition."

The family tradition started in 1965 continues at Hoober Inc., as much of Charles "Bud" Hoober's descendants continue to work selling red equipment. For families like the Hoobers across the nation, the new technology offered by the Axial-Flow offered better technology and opportunities to grow farms as well as businesses.

1980 Grain Embargo

▲ In 1980, the Carter administration imposed a grain embargo on the Soviet Union. But the Russians were able to get grain from other sources and the lowered demand pushed down prices on the United States' bountiful 1980 harvest. *Wisconsin Historical Society*

Field Work

◀ The Axial-Flow had its work cut out for it into the 1980s. The down market meant only the most successful farmers could afford the new technology. Harvester was failing as well. The machine proved a bright spot in dark times. *Wisconsin Historical Society*

INTERNATIONAL HARVESTER 1440

When the Axial-Flow combines launched in 1977, IH had only 300 machines to offer. This meant that the company had to pitch the new machines carefully, letting people know it had a wonderful new technology but also selling the 15 Series machines. The fact was IH didn't have enough Axial-Flow combines to go around.

Dealer presentations were set up in Champaign, Illinois, and Des Moines, Iowa. A more experienced salesperson was assigned the duty of selling the 15 Series machines, while one of the younger guys was tagged to bring out the new Axial-Flows.

Gerry Salzman was that younger guy.

"I'm thinking we were allocated 60 of the first 300," Salzman said. "We followed these machines very closely."

One of those first 300 went to Chet and Dan Eyer. The father and son still recall the first time they were shown an Axial-Flow combine, back in 1977. The IH engineer showing them the machine was trying to explain its efficiency.

"'This combine does such a good job that a chicken would starve if it had to live on what fell out of the back of it,'" Dan recalled being told. The engineer went on to say the combine could harvest corn at 2 to 3 miles per hour—quick for the time.

1977 International 1440

▶ This illustration shows a brand-new 1440 in 1977. The combine was released as part of a limited production run of 300 Axial-Flows in 1977. It entered full production in 1978. *Case IH*

Dan was duly impressed: "I'm 22 years old and I'm listening to this guy and I'm going, 'Man, Dad . . . we need this thing."

Chet agreed, and the family ended up with one of the first 300 machines. Which is how they ended up meeting Salzman.

All of the regional sales reps kept very close tabs on those first 300 machines. "We worked with service and parts guys to make sure people were trained and parts were available," Salzman said. "And we paid special attention to any calls that came in when a customer had an issue."

The reps also gave surveys to the 300 owners. The surveys were done personally—the rep would ride in the combine with the customer and ask questions about how the machine was performing as the customer drove.

"The survey started in late September 1977," Salzman said. "I took all the Illinois guys, and another rep took the Iowa guys."

One of those Illinois guys was Chet Eyer.

Salzman climbed into the combine's buddy seat next to Chet and started asking questions and writing answers on his clipboard.

They made one round.

"Gerry," Chet said, "You know what, I'm tired of driving this thing. You drive and I'm going to ask you some questions."

Salzman agreed, and he recalls the questions Chet asked were direct and not always easy to answer. The two hit it off after that.

"He's the one that . . . enthused me on this combine. Gerry was probably one of the easiest guys to talk to I ever met," Chet said. "Gerry and I have been good friends ever since."

"And I tell you what I liked about the combines the most: we finally got rid of those straw walkers," Chet said.

The change from a conventional combine required new techniques. For the Eyers, just as with other farmers, they had to figure out how to set it up, which took some time. But once it was set up, the machine was a keeper.

"The sample was beautiful, so it was quite amazing," Dan recalled. "We bought another one the next year and a couple neighbors came by and they bought one the next year.

"We kept that [combine] until we bought a 1660. We kept it that long because that thing had no leaks, nothing go wrong with it. It was built perfectly."

International 1440

▲ The 1440 was a Class 4 machine and featured a 135-horsepower 436-ci engine and 145-ci grain tank.

Wisconsin Historical Society #114748

Easy Service

▶ The new Axial-Flows had this access panel to make service simpler. *Wisconsin Historical Society #117406*

International 1440

▶ Creating the Axial-Flow combine took more than a million man hours of research and development. This image is of a 1978 model. *Wisconsin Historical Society*

International 1440

▼ Corn is where Axial-Flow technology shone the brightest early in development. Exceptionally low grain loss and crackage the machine returned in corn, as well as other grains, were strengths of the Axial-Flow models from the very beginning. *Wisconsin Historical Society #114834*

THE CAB

1400 Series Cab

▶ The cab for the Axial-Flow line was state-of-the-art. The interior was 58 inches wide—about 18 inches wider than the competition at that time. The cab was also exceptionally quiet and could be equipped with a radio or eight-track player, interior lighting, and a cigar lighter. *Wisconsin Historical Society #114832*

1400 Series Control Panel

◄ The main control panel included warning lights, a gauge cluster, and state-of-the-art controllers for header height, reel speed, and reel lift. Three broad speed ranges could be selected, and hydrostatic drive allowed infinitely variable speed adjustments. *Wisconsin Historical Society #114572*

1400 Series Interior

▲ The chair was six-way adjustable. The machine also featured hydrostatic steering and an adjustable steering wheel. The interior air was heavily filtered, and the cab also featured heat and air conditioning. *Wisconsin Historical Society*

Visibility

▲ The 1400 Series cab offered terrific visibility, as shown by this image of a combine picking up windrowed grain. *Wisconsin Historical Society #117326*

INTERNATIONAL HARVESTER 1460

Gerald Hergott started working at his uncle's IH and Studebaker dealership in 1960, when he was 18 years old. He eventually partnered with Murray Hergott to run the dealership, and they are now turning it over to Gerald's son, Lenn. The Hergott Dealership is a single, family-owned store, and one of the largest in Canada.

The first Axial-Flow combine Gerald saw was a prototype that was trucked up to Saskatchewan in the late 1970s.

He and a few trusted customers inspected the combine, which sat out in a field behind St. Peter's School. When their brief inspection was done, the combine was packed away. "They had it all padlocked," Gerald said. "Even I couldn't really see what was going on—nor did I really want to know. Then it got shipped off to Australia."

Roughly a year later, the Hergott dealership received a 1460 and a 1480. They held a demonstration and invited both regular IH customers and others who were running competitors' combines.

After the show, Gerald and his salespeople went on the road to visit the customers who had attended the demonstration. That week, they sold six Axial-Flow combines to people who owned John Deere combines.

Len and Gerald also used the machines on their own farm, and their performance at home helped them sell the machines to their customers.

1977 International 1460

▶ This is the very first 1460 produced in 1977. The machine was tracked down by Case IH for an Axial-Flow anniversary and is owned by Matt Frey. The machine is put to work each year. Despite its pedigree, this 1460 is no garage queen. *Lee Klancher*

International 1460

▲ The 1460 appeared in 1977 as part of a run of 300 units and came into regular production in 1978. The 1460 was a Class 5 combine that used the same rotor and chassis as the 1440, but featured a more powerful 170-hp engine and a larger 180-bushel grain tank. *Wisconsin Historical Society #114728*

1977 International 1460

The first 1460 in the field with International Harvester field technician Jerome Ripperda, who recalled drilling a service hole in an early machine to remove a shaft. His hand-drilled hole became a factory retrofit, as it was the only way to service the shaft. *Lee Klancher*

International 1460

◀ This 1460 is fitted with a 1063 six-row corn head. The heads available for the original 1400 combines were the 800 Series heads, which were available as four-, five-, six-, and eight-row models. This 1460 is owned by Daniel J. Tordai. *Lee Klancher*

International 1460 Rice Special

▼ In late 1978, International introduced a rice-harvesting version of the 1460. The package included a special rice rotor, larger heavily lugged tires, and optional tracks. *Wisconsin Historical Society #117310*

International 1460

The 1400 Series featured quick-disconnect headers that allowed the operator to lower the head, flip some latches, and back out of the connection.

Wisconsin Historical Society #114835

1982 International 1460

◀ Photographed in 1982, this 1460 was fitted with a 963 six-row head. During the machine's production from 1977 to 1985, a number of improvements were incorporated. The engines were upgraded, a new alternator was introduced in 1979, and a spin-on fuel filter and rock trap were added, among many other improvements.

Wisconsin Historical Society

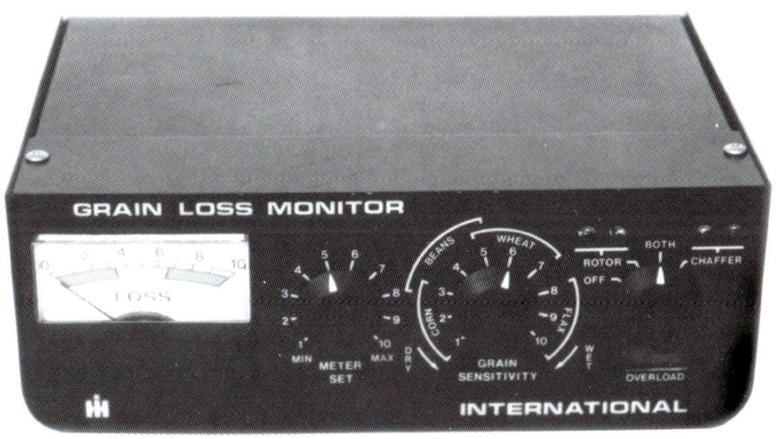

Grain Loss Monitor

◀ An optional grain-loss monitor was offered on Axial-Flow combines beginning in 1980. This model built prior to 1980 was either preproduction or experimental, as it looks radically different than the production version. In the production version, six sensors monitor shaft and oscillating speeds. If any shaft slows or stops, an alarm sounds, warning the operator something is most likely plugged. The six shafts are the cleaning fan, shoe tailings elevator, clean-grain elevator, rotary air screen, and discharge beater. *Wisconsin Historical Society #114854*

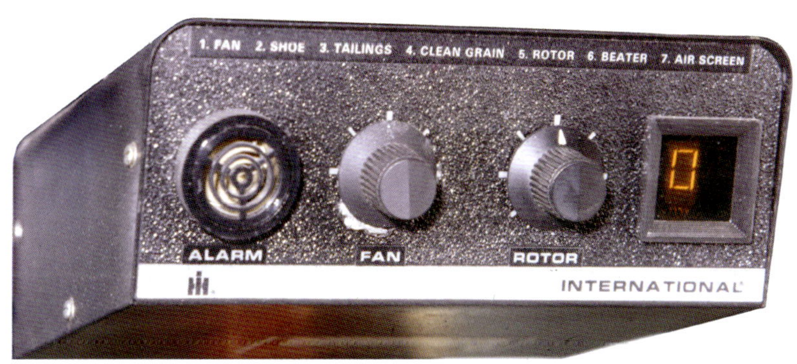

Grain-Loss Monitor

▶ This grain-loss monitor also differs slightly from the 1980 production version.

Wisconsin Historical Society #114874

Rice Harvesting

▶ Perfecting rice harvesting with the Axial-Flow was a painstaking process that required a lot of experimentation with different protrusions on the rotor. *Wisconsin Historical Society #117310*

Axial-Flow Combine Sketch

◀ While technology was the star of the new Axial-Flow, the design was also groundbreaking. Harvester industrial designer Jerry McGirk drew a machine with smoother lines and a more integrated, logical layout than anything on the market at the time. *Gregg Montgomery Collection*

AXIAL-FLOW DESIGN AND STYLING

By Gregg Montgomery

The IH 1400 Series Axial-Flow combine had the first rotary threshing system manufactured and brought to full production. While there had been prior attempts to build such a system, none was ever successful on a production basis.

A combine with such revolutionary engineering deserved equally revolutionary design and styling. Thanks to the IH Industrial Design Department and lead designer Jerry McGirk, that is exactly what it got. Until the introduction of the first IH Axial-Flow design, all combines were pretty much a series of flat sheet-metal boxes enclosing various components and stacked on one another. There was little or no thought given to the overall appearance or the integration of various components into a singular form that bespoke the machine's functionality.

After producing numerous concept sketches and scale models for the Axial-Flow design, a quarter-scale production layout of the selected concept was begun. In the 1970s before computer-aided design, this was a chore. The layouts were produced on a 12x8-foot vertical drafting board, a task that was physically as well as mentally taxing.

The new combine was given a dynamic profile that reflected the mounting angle of the rotary separator and gave the entire machine a clean and aggressive look. The original Axial-Flow set the design trend for combine aesthetics for the next 40 years. One can still see IH's engineering and design influence in virtually every combine built today. The Axial-Flow combine was—and is—a truly revolutionary product that has transcended the years both functionally and aesthetically.

INTERNATIONAL HARVESTER 1480

The first 1480 was produced in 1978, and Don Murray's memoirs show that 308 machines were shipped that year. He reported that the estimated engineering expense to develop the 1480 was $7.89 million, going back to the first CX-14 experimental built in 1968.

In 1979, Englishman Will Bushell flew to the United States and traveled to the dusty little town of Vernon, Texas. He had spoken on the phone with a man who told him he had a job in America for him.

But when Bushell showed up in Vernon, no one was there to meet him. After a long wait, a big red pickup truck screamed up and a broad Texan jumped out with a six-pack and a smile.

International 1480

▼ The big 1480 was introduced late in 1978 and was the largest machine on the market at that time. *Wisconsin Historical Society #114831*

Bushel spent the next six months busting his hump in a custom-cutting outfit, following the harvest from Texas to South Dakota and all the way north to Wolf Point, Montana, and then back south to Texas. They finished in December 1979.

"We learned to run combines in our sleep," Bushell said. "We worked horrendous hours when the weather was good. When it isn't you do your laundry and maybe go to town and find a bath."

His experience running 1480 combines while custom-cutting made a lasting impression.

"It's one thing to field-test a machine from an engineering perspective," Bushell said. "Give it to a customer who runs it for thousands of hours and things turn out differently."

He also saw the immediate positive effect the Axial-Flow had on output. "There was less cracked grains and it was kinder on the sample," Bushell said. "In those days you would get a premium price for better samples."

When Bushell returned to England, he took a job with International Harvester Great Britain. He's worked with red combines ever since, launching new products around the world. Like Cam Beert, he knew the combines intimately and worked with customers as well as staff to get the machines working properly.

Back in Europe, he found that Harvester had a big challenge helping its customers understand the advantages of rotary combines.

"The biggest thing was, it was a radical change," he explained. "Everybody had been used to the development of conventional combines, conventional cylinders. We were selling this single rotor, and what you got to remember is that for any market, when you go with something that was as revolutionary as the Axial-Flow was in those days, you were basically asking the customer, who has spent months of the year cultivating, planting, nourishing the crop to gain an end product that he is then going to sell, which is going to be his livelihood . . . you are asking him effectively to risk all of that by using a machine that is completely different to what all the other competitors are selling."

What worked was showing the customer exactly how the combine would perform.

"Pretty early on, we had to take a combine anywhere, to anyone that we could get one and prove it in the field, and run it alongside their existing machines," Bushell said.

A key was helping new operators properly set up the machines.

"On a conventional combine, if you want to clean the sample up, you would close down the sieve, but decrease the wind of the machine, whereas with an Axial-Flow, you did it completely the opposite. . . . For some people, they just couldn't comprehend that. You set the machine up in many occasions, I'd go out and I'd set the Axial-Flow up and it'd be going great.

"You go back the next day or a couple of days later, and the guy shut it all back down like it was a conventional combine again, and so he was pouring grain over the back or not getting the right sample, or not able to get the throughput because he got everything clamped down tight. That was a very big challenge. It was almost a cultural challenge, if you like."

The American and the European sales teams faced these challenges with customers in Des Moines, Iowa, Seiches-sur-le-Loir, France, and Mukinbudin, Australia.

Harvesting with rotary combines required a shift of mindset for farmers around the world.

International 1480

▲ The big Class 6 machine featured a larger 30-inch rotor driven by a 190-hp version of the 436-ci engine. The 208-bushel grain tank was larger as well. *Wisconsin Historical Society*

International 1480 Rice Special

▲ The 1480 Rice Special featured the rice rotor, heavier tires or optional tracks, a longer unloading auger, and a 466-ci engine. Rice is a noted power hog, and the larger motor improved performance with the heavy, wet crop. *Wisconsin Historical Society #114845*

International 1480

▶ The 1480 could drive a large, eight-row 883 or 884 head. The 883 was for 28- or 30-inch rows, while the 884 was designed for 36-, 38-, or 40-inch rows.

Wisconsin Historical Society

International 1480 with 154-Inch Belt Pickup Head

◀ International Harvester offered a variety of pickup heads for the 1400 Series. The head above is a brand-new 154-inch version harvesting wheat near Devil's Lake, Manitoba, on August 26, 1982. The customer was one Mr. Hartman, and notes indicate his dealer was located about 4 miles south in Langdon, North Dakota.

Wisconsin Historical Society #114938

RAT PATROL: THE 1979 STRIKE

By Lee Klancher

Jamie Horst was one of the best in all of International Harvester at motivating people to build quality machines. His programs at the East Moline Plant were popular with employees and copied in plants around the country. When union labor went on strike in 1979, those skills became of limited value.

In fact, he and any other non-union workers still at work at IH—everyone from management to janitors—were put to work building machines.

Jamie was asked to cut pieces of metal on a sheetmetal break. To save time, he tried to cut three pieces at once.

He broke the machine. The entire line was stopped.

The line supervisor came out.

"I don't know what happened," Jamie explained. "The machine stopped."

He was demoted to rat patrol.

The struggles with labor at IH were not unique. In the late 1970s, nearly everyone involved in the heavy equipment industry was struggling to deal with organized labor. The times were hard for equipment manufacturers, and the unions wielded tremendous power.

Under the leadership of Archie McCardell, IH management's differences with the United Auto Workers (UAW) became cataclysmic. McCardell instituted wide-ranging cost cuts, including firing most everyone in the company who had experience negotiating with labor. McCardell and human resources vice president W. Grant Chandler personally directed union negotiations. In August 1979, the pair decided to go after several relatively minor points, including mandatory overtime.

The inexperienced duo bungled the process and the situation deteriorated quickly. Both Deere & Co. and Caterpillar suffered short strikes but were able to settle with the UAW in October. McCardell and Grant, however, demanded mandatory overtime and refused to concede to the union's demands.

The UAW, bolstered by high reported earnings at IH that fall, also refused to give in, and on November 1, 1979, 35,000 UAW workers from 21 IH plants went on strike. Thirty-six percent of the IH workforce sat idle.

Hard Days

▲ For 172 days in late 1979 and early 1980, 36 percent of the IH workforce went on strike. CEO Archie McCardell's administration took over union negotiation and handled it poorly, and the result was a long, bitter battle that benefited no one.

Wisconsin Historical Society #56753

At the East Moline plant, which was staffed mostly by UAW workers, the strike led to chaos. The picket lines were huge, and the massive parking lot at East Moline became so full that the few people still working at the plant had to come in early just to get a spot.

"You go through the picket lines every day, which was not fun because the guys we knew, we knew to be good guys, would harass you, would pound on your hoods," Horst said.

The strikers demanded to see IDs. "Show us your badge," they would say.

Inside the plant, people were sabotaging things when they could, kicking over corn heads and doing anything else that would slow down the work.

That's where rat patrol came in.

"Rat patrol is where they assigned people to work with the guards and watch for sabotage," Jamie said. Thanks to his shoddy work on the sheetmetal break, he spent plenty of time patrolling the grounds.

As the strike dragged on, relations with the plant workers deteriorated.

"The longer it went," Jamie said, "the tougher we knew it was going to be to reassemble the workforce and to get them back to work."

The strike finally settled in April 1980 after a record-setting 172 days.

Steve Horst, Jamie's brother and an industrial engineer at the plant, estimated that it took at least six months for the plant to return to anything resembling a healthy, stable work environment.

IH won no concessions from the UAW.

"At the end of the day, nobody came out ahead," Steve said. "Everybody lost."

By the time the strike settled, the farm economy was taking the hardest fall since the Great Depression. Prices dropped precipitously, interest rates rose above 20 percent, and farmers defaulted on loans and went bust.

The hard times, combined with IH's debt load, led to cuts. IH laid off workers throughout the early 1980s. "That next four years was brutal," Steve said.

"We had a blood bath," Jamie agreed.

The layoffs were hard on staff. Union rules required senior members be retained, which often meant that critical workers had to be let go.

"It was absolutely a nightmare," Steve said. "You've got people that knew location, knew where to take materials, knew how to load a crane, knew how to drive a forklift, knew how to drive a whatever, that are now gone. This went on and on and on."

The timing for the red combine was equally brutal. The East Moline plant was changing over to build the new Axial-Flow, which was enough of a challenge without half the workforce being let go.

In 1984, it came to light that Tenneco was slated to purchase International Harvester's agricultural division. For some IH employees this came as a relief, as it meant the company would survive rather than dissolve.

This was not the case at the East Moline plant.

"Word on the street was that the East Moline plant was done," Jamie said.

Steve and Jamie remember watching a combine be driven out of their plant and down the street to the Farmall plant to see if it would fit. The Farmall plant had recently been modernized as well.

As far as they were concerned, the writing was on the wall. When they were told that a big announcement was coming, they believed the Farmall plant would take over combine production and East Moline would be shuttered.

"Now, after going through all of the morale issues that we had leading up to this, now we're lower than a snake's belly in a wagon track because we know it's over," Steve recalled. "We're just waiting to see how much time we have."

Steve was forty years old at the time, with two young children. A historically bad farm economy and rough times for heavy equipment manufacturers meant other jobs would be hard to find.

The big announcement was made not by management but through a slip when the television news announced that tractor production was moving to Racine, Wisconsin. The newer, high-technology Farmall plant in Rock Island, Illinois, would close.

Steve believes East Moline survived because they were extremely cost effective. The Farmall plant was modern, but it needed to move a huge volume of machinery to be profitable. The farm economy in 1985 was horrid and a turnaround was not expected anytime soon. The bottom line may well have been the dollar.

"To this day, I still believe, and this is my thought, that East Moline stayed the course because of it being the age that it was," Steve said. "I feel that when they finally pushed the numbers, the number crunchers and the accountants said, 'Wait a minute, this plant over here is making us money.'"

The old guard at East Moline survived, and red combines would be built there for nearly two decades.

INTERNATIONAL HARVESTER 1482 PULL-TYPE COMBINE

The future of the pull-type combine at Harvester was a hotly debated topic. While the bulk of the Harvester team believed the new pull-type would be based on the 1460 combine, Dathan Kerber, the 1482 project engineer, pushed for a different approach.

Kerber argued that the farmers in the heart of the market for the pull-type (the Dakotas, eastern Montana, and the Canadian prairie) were moving up and purchasing larger tractors and would desire a large pull-behind machine based on the 1480.

He wasn't alone in his belief. At the time, farmers working big acreages tended to put most of their equipment investment into large, expensive four-wheel-drive tractors.

These farmers preferred to use the big machines to operate big pull-behind combines. The pull-behinds were a bit more economical than the self-propelled machines.

As word spread that the Axial-Flow combines were more efficient than conventional machines, customers and dealers in the central part of the United States and Canada were clamoring for an Axial-Flow pull-behind machine.

The cost to develop such a large machine was as hefty as the tractor required to run it, and it required considerable effort to convince the management to tackle the project.

Once the decision was made, the task proved more difficult and expensive than expected. An all-new frame, power take-off (PTO), main-gearcase drive, and electronic/electric hydraulic controls were engineered and tested at a cost of nearly $3.5 million.

International 1482 and 4166

▼ Introduced In 1980, the 1482 was far and away the largest pull-behind combine on the market and it came at a time when an upgrade of Harvester's pull-behind model was long overdue. This one, as well as the 4166 tractor, is owned by Darius Harms, and was photographed on the former Air Force base in Rantoul, Illinois, that serves as the grounds for the Half Century of Progress tractor show.

Lee Klancher

The result was a machine that raised the bar for the entire industry. The 1480 had a 78 percent greater capacity than the 914 it replaced. "I doubt if anyone has equaled it overall," Murray wrote. He also stated the machine's rating for 130-hp tractors was conservative, and estimated the machine could handle up to 200-hp tractors.

International 1482 Prototype

▼ The 1482 used the 30-inch rotor from the 1480 combine and had a 245-bushel grain tank. The big 1482 was PTO-driven and required a tractor with at least 130 hp to operate it. This 1482 pull-type combine was photographed harvesting wheat in the Dakotas in early fall 1978. *Wisconsin Historical Society #114839*

International 1482

▶ The 1482 was available with a 17.5- or 12.5-foot header or a 132-inch draper pickup. This 1482 pull-type combine is seen behind a 1086 tractor and was photographed in 1979.

Wisconsin Historical Society #114838

International 1482

◀ Hydraulics remotely controlled the header height, reel height, and the unloading auger. This 1482 pull-type combine was pulled through barley windrows by a 5488 tractor with duals. Image taken August 19, 1982, near Grand Forks, North Dakota. *Wisconsin Historical Society #114837*

INTERNATIONAL HARVESTER 1470 HILLSIDE COMBINE

The Hillside machine was produced by adapting the 1460 chassis with four-wheel drive and the ability to handle a 44 to 49 percent side slope. According to Don Murray's papers, the machine's size would have made it difficult to move to market in the Pacific Northwest. As a result, the team looked to hire a third party to do much of the engineering and the final assembly of the 1470. IH selected RAHCO in Spokane, Washington, owned by Raymond A. Hanson, the inventor of the combine's self-leveling control system.

RAHCO was based in an old World War II plant. Hanson was semiretired and his sons ran the place. The company had a solid engineering team, but Harvester had concerns about RAHCO's ability to manufacture the machines.

Despite difficulties with development and some fundamental differences between RAHCO and IH, the 1470 Hillside Combine was introduced at a big product announcement held at a convention center in Spokane on June 19, 1979. Murray described the introduction as "premature." The price announced at the event was $115,000—astronomical considering that the 453 sold for $50,732 in 1976. The machine was not released, and engineering went back to work attempting to reduce the price and refine the final product.

A half-dozen Harvester engineers, including Leroy Pickett and Rich McMillen, were brought in to rework critical areas

International 1470

▶ Harvester led the market in hillside combines with the four-way-leveling 453, and it maintained that market position with the 1470 Axial-Flow hillside combine. This 1470 Hillside combine is working wheat in Idaho in 1979. *Wisconsin Historical Society #114879*

International 1470

Power to the 1470 was supplied by a 210-hp DT466 engine. The 1470 was based on the 1460 and used the 24-inch rotor from the 1460 and a 145-bushel grain tank. The four-way-leveling machine could handle slopes with a grade of up to 48 percent.

Wisconsin Historical Society #114884

International 1470

▶ The 1470 started life as a modified 1460 built at East Moline that was shipped west to the Raymond A. Hanson Co. (RAHCO) in Spokane, Washington. RAHCO had a good reputation for engineering but turned out to be less than stellar at manufacturing—at least early in the process. Once the bugs were worked out, the 1470 was introduced and proved a good machine. This 1470 Hillside combine is working Idaho wheat in 1979. *Wisconsin Historical Society #114875* of the combine. The hydrostatic four-wheel-drive system and related electronic controls proved particularly problematic.

By early 1980, the bugs were worked out and the 1470 was released to production. The engineering cost was about $1 million, about double the estimate. Murray reported sales of 102 units in 1980 and another 145 in 1981—solid sales for the time, but not surprising to Murray.

"The 1470 hillside really didn't have a worthy competitor from the standpoint of capability," he wrote. Which was good—the retail price in 1981 was $131,629—more than $15,000 higher than the John Deere 6622 hillside, and $25,000 more than the Massey-Harris hillside combine.

When Don Murray retired from Harvester in 1982, he was presented with a photograph of four 1470s working Kentucky bluegrass just southeast of Spokane. Cam Beert inscribed the back, "A family of four brothers owned eight John Deeres. They traded them in for four IH combines. It is well-known that one 1470 will do the work of two John Deeres."

Hillside Combine in Action

▲ A 1470 combine working in the field. *Wisconsin Historical Society*

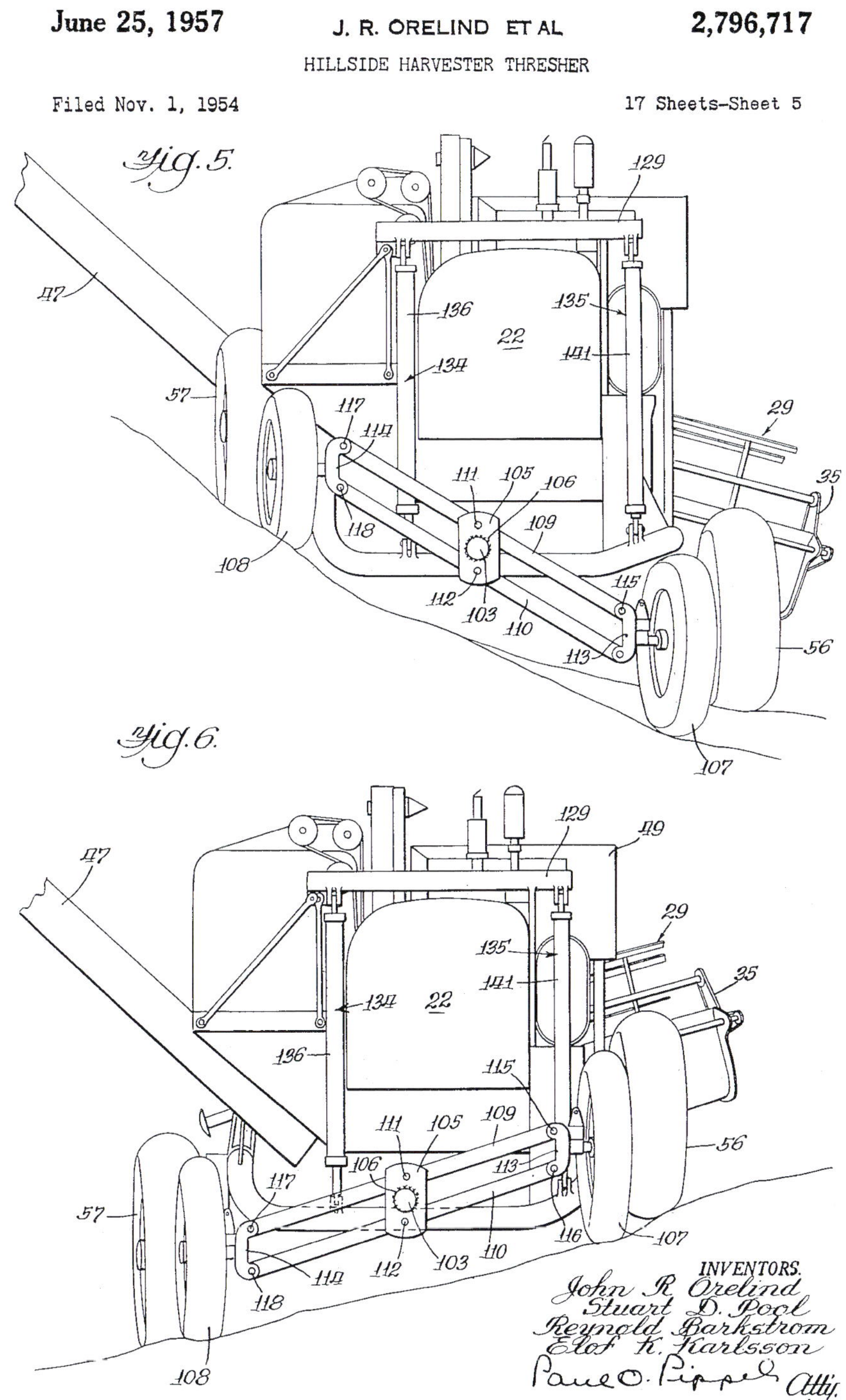

1957 Hillside Harvester Patent

▲ The influence of International Harvester's 1950s engineering team lasted into the 1970s. This patent co-authored by Elof Karlsson and other Harvester engineers is one of the most important patents of the hillside machine and is referenced by 16 other patents filed by Deere and Company, Massey Ferguson, New Holland, and AGCO. The last reference to this patent was in September 1991. *U.S. Patent Office*

INTERNATIONAL HARVESTER 1420

The 1420 was ready for production on March 1, 1979, and intended to be released early in 1980. Unfortunately for the struggling company, the leadership's battles with organized labor sparked a massive strike on November 1, 1979. A compromise wasn't reached until late spring 1980 and International's factories were run with skeleton crews unable to ramp up for a new machine.

First production of the 1420 occurred in 1981. The little machine replaced the popular 715, but did so at a much higher retail price. The 715 was priced at $33,690 in 1977, and the 1420 came in at $64,443 when announced in 1980.

"We undoubtedly lost many of the lower-income, full-time farmers," Murray wrote. "It was easy for them to get good used machines, or lower-priced competitive machines with less features."

Murray reported retail sales of 965 machines in 1980, with another 965 carried over into inventory for 1981. Once the factory was back in production, they produced double the volume needed. This was true across the board for Harvester in 1981, a year that would prove one of the grimmest ever for the company.

The 1420 was—and is—a wonderful little machine for smaller-acreage farmers. The market at that time, however, was the worst in recorded history since the Great Depression of 1933.

International 1420

▶ The last 1400 Series model was the little 1420 introduced in 1981. Powered by a 112-hp 358-ci engine, the 1420 had a 20-inch rotor and a 125-bushel grain tank. The lighting system on the 1400 Series was powerful for the day.

Lee Klancher

International 1420

▶ The 1420 was designed to replace the popular 715 and was the first Harvester combine to feature Electro Hydraulic controls, meaning the hydraulics were controlled with switches and solenoids.

Wisconsin Historical Society #114811

International 1420

▶ The Axial-Flow combines featured fewer moving parts than straw walker combines and offered greatly improved reliability. The single rotor alone replaced 16 moving parts in the 15 Series models. Access was also simpler: the drives for the 1420, like the rest of the 1400 Series, were accessed by this flip-up side panel. *Wisconsin Historical Society #114754*

International 1420

◀ The unloading auger and tube could be controlled from the cab. Unloading speed was up on the 1400 Series, moving about two bushels of grain per second, meaning a 180-bushel tank could be unloaded in about 90 seconds. *Wisconsin Historical Society #114755*

HOW AXIAL-FLOW SEPARATION WORKS

"Rotary is like slinging a bucket of water. The water will stay on the bottom. The harder you sling it, the harder she'll press against the bucket bottom."
—IH field test engineer Camiel Beert

Axial-Flow Combine

▲ The Axial-Flow combine uses a long rotor with stationary concaves and grates underneath to separate grain. The crop is fed into the front of the machine by rotating beaters—no differently than the older combines. The difference comes after the grain comes into the machine. The material is fed into the rotor and spun by the rotor and forced air, with centrifugal force pushing the grain out through the bottom of the rotor tube. *Wisconsin Historical Society #114866*

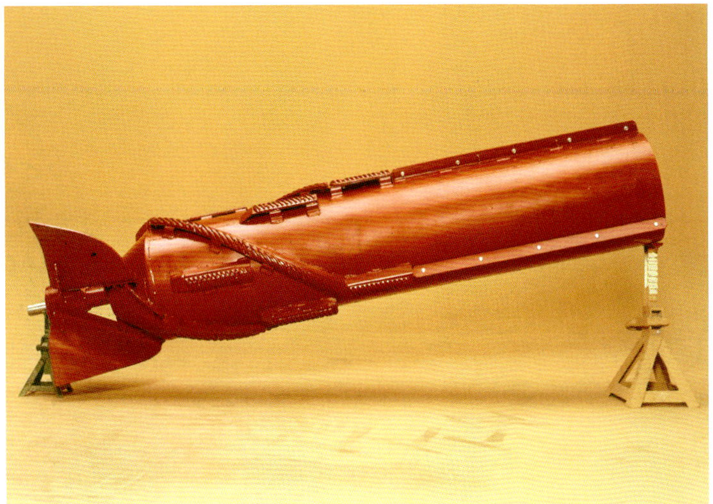

Axial-Flow Rotor

▲ The design of the rotor is deceptively simple. Refining the precise shape and protrusions required a decade of trial and error. Test engineer Red Gochanour remembers welding different-shaped pieces on the rotors in sheds all over the country. This image of a preproduction 1420 rotor was taken on June 22, 1979.

Wisconsin Historical Society #114751

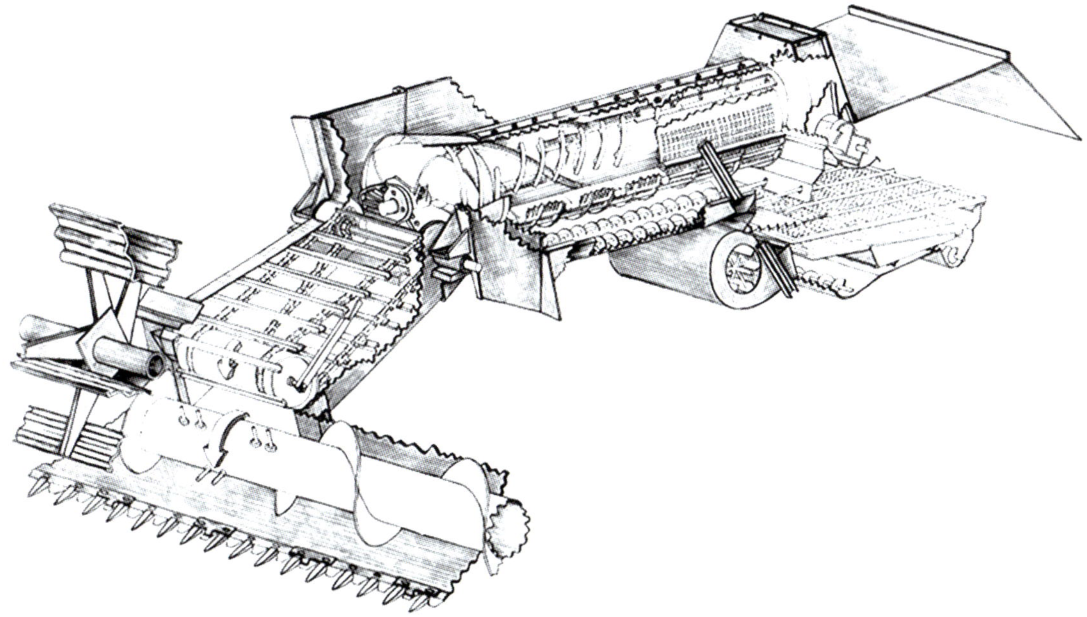

Material Flow

▲ This drawing shows how material flows into the front of the combine through the rotor, and how grain separates out the grates while the chaff is blown out the back. *Case IH*

Axial-Flow Concaves and Rotor Tube

▲ The concaves are where the material enters the rotor tube, and the design of that shape was another item that required thousands of hours of testing and design.

Wisconsin Historical Society #114731

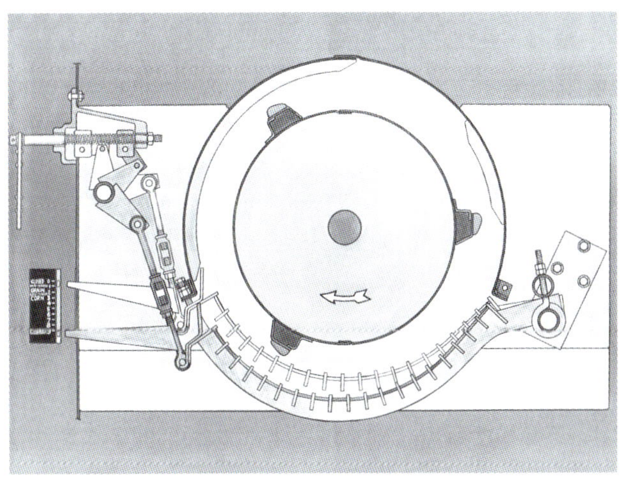

The Concaves under the Rotor

▲ This drawing looks at the rotor from the end and shows the concaves below it. They are adjustable—getting these settings right is part of the art of running a combine. *Case IH*

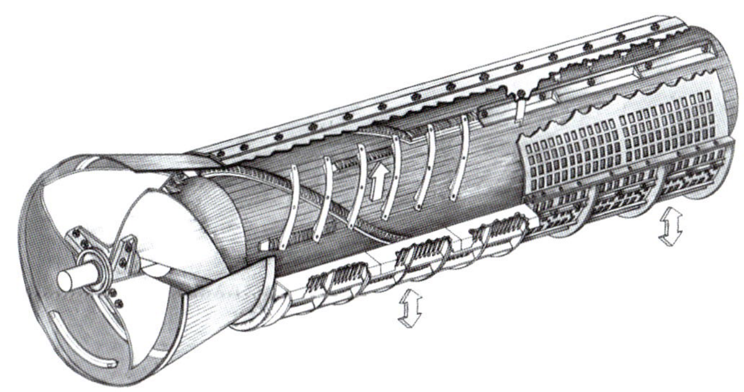

The Rotor and Grates

▲ This view shows the rotor and where the separated material goes. *Case IH*

THE PROBINE

In 1979, Harvester was investing in new technology at an unprecedented rate. Under the leadership of former Xerox executive Archie McCardell, money was poured into research and development. The effort was an expensive gamble. Harvester was behind on technology at the time, and any path to a sustainable future had to include upgrading the product line. Many vilify McCardell for spending so recklessly when Harvester was saddled with such immense debt. But in the end, his emphasis on research investment produced the technology that would become the Magnum tractor. That was critical to the long-term survival of red machinery. Other ventures, however, proved less fruitful.

One of the most fascinating of these failed efforts was the protein harvester developed by the Advanced Harvesting Systems venture group. AHS was formed as an independent company and staffed with a mix of scientists, engineers, designers, and visionaries. The idea was to create an all-new method of harvesting and utilizing crops.

One can imagine the group looking back on the success of the Axial-Flow combine and hoping to re-create some of that with this group. Certainly some of the most successful engineering ventures in history have involved small groups left relatively unmolested to create something new. The original Farmall, the John Deere New Generation tractor, and even the Axial-Flow combine are good examples of situations in which a small team of people relatively isolated from management (or protected by management) were able to make great strides.

The AHS group was one of three new IH venture groups set up as independent companies. AHS was under the direct

Protein-Harvesting Patent

▶ This patent is for a method of harvesting that shreds the crop into small particles and presses it to separate protein from the crop. The patents were filed filed by International Harvester engineer George Cicci in April 1982. The first patent appears to have expired, and the second was held by Navistar into the mid-1990s. *U.S. Patent Office*

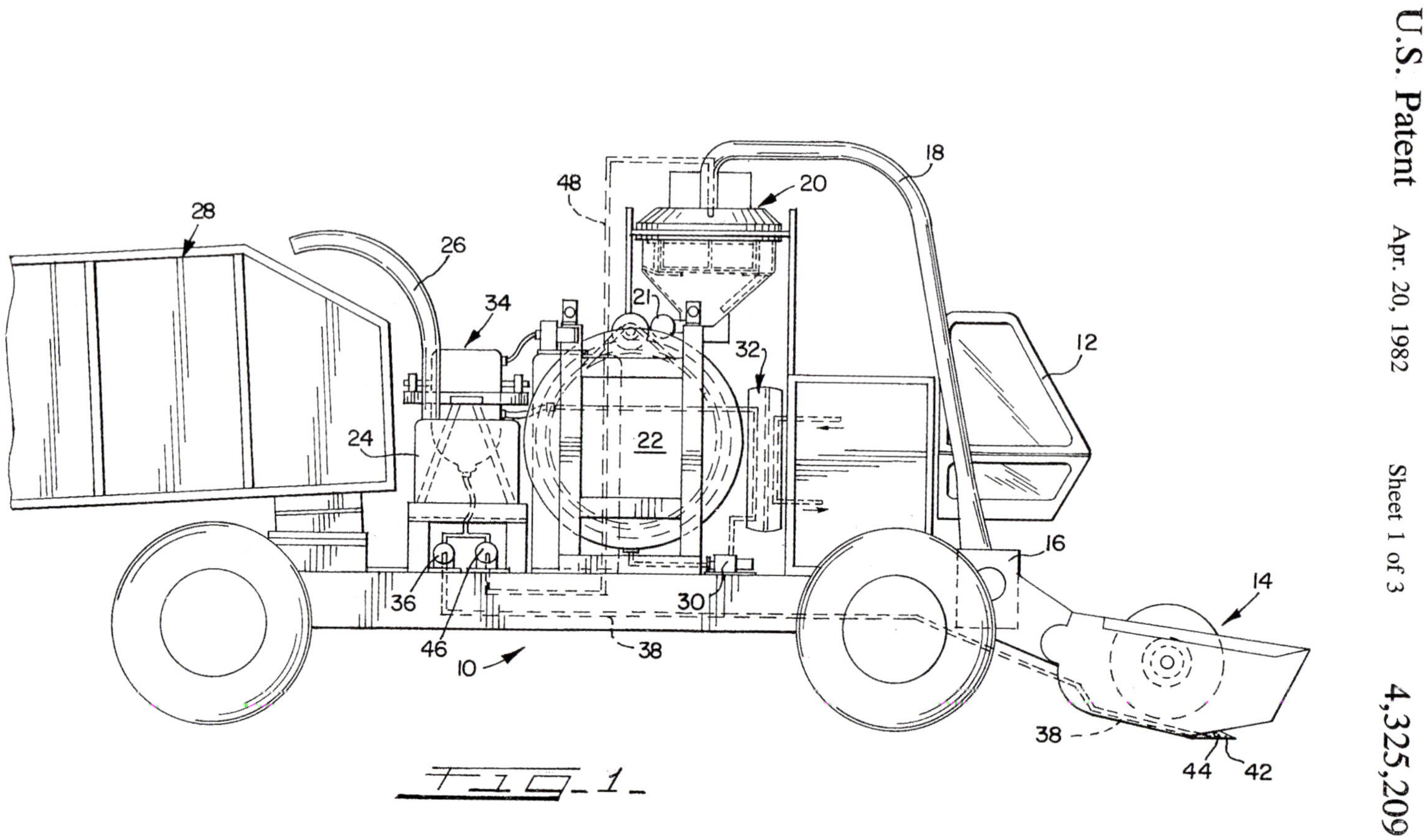

Advanced Harvesting System

▼ In the late 1970s, IH assembled a group of scientists, engineers, and others from within and outside of the company to develop a protein harvesting machine. This is a concept sketch of one of the early designs. *Gregg Montgomery Collection*

management of one of McCardell's trusted leaders, Bob Potter. The concept it explored was the creation of a machine that would extract protein from green crops, and do so right in the field. The technique was called "protein harvesting," and the machine it designed was known as the Probine.

Rick Tolman had been designing forage-harvesting and haymaking equipment for the Gehl Company. He was recruited to join the team for his experience with green crops. He explained that the machine they were developing would process alfalfa in one pass and separate it into three components: green solid (similar to silage), green juice, and brown juice. The green juice was spun in a centrifuge to create a protein product that could be used in a variety of applications as animal feed or (with further processing) to feed humans. The brown juice was sprayed back onto the ground as fertilizer.

Far-out stuff.

Gregg Montgomery was hired as the project's industrial designer in 1979. He was attracted to the independence and creativity of the project. "I hired in and I think I was the fifth or sixth hire," he recalled. "Ended up with about 50 people when it was all said and done."

AX-3 Harvester Concept Sketch

▶ This later concept shows a pull-type two-piece harvester with a front processing unit and rear storage wagon for green fiber, and a storage tank for protein concentrate. A version of this unit was prototyped and field-tested in the Imperial Valley of Southern California.

Gregg Montgomery Collection

The mix of talent on the project was impressive. "They had scientists, they had marketing people, they had agronomists," Montgomery said. "They had animal people that studied cattle because one of the thoughts on this was that you could cut a crop like alfalfa, bring it into this machine, extract protein from the crop that typically would just be bypassed in a cattle system, and therefore not get used because they really didn't use all the protein that was available in the crop."

The project proceeded to the point where the group had working prototypes that still required a considerable amount of development to finish. Potter intended to continue to

Probine Prototype

◀ Several working prototypes were built and tested. The first was a stationary unit. This field-going prototype shown testing in California had a tendency to plug up its macerating device. Program manager Bob Alverson is shown on the unit during a teardown review.

Gregg Montgomery Collection

AX-7 Probine Concept Rendering

◀ This was the largest concept machine envisioned and was the one being worked on when the Advanced Harvesting Systems (AHS) project was shut down. It had several advanced features, including its "hawk-wing" doors—similar to those on the 2015 Tesla Model X that appeared 34 years later. Other AX-7 features included an electronic periscope for monitoring the wagon and chopper chute, a 6-foot-diameter onboard cone press, and an onboard centrifuge for separating the protein.

Gregg Montgomery Collection

invest—Montgomery recalls drawings of a large, elaborate facility intended to be constructed at Burr Ridge, Illinois, in part for the AHS group.

In late 1981, Montgomery was tapped to head up the IH Industrial Design group to work on the Harvester line of combines and the full line of 50 Series–style tractors. One can safely assume funding for the Advanced Harvesting Systems group became problematic in the last days of Harvester—money was tight for everyone in 1983 and 1984, and thousands of employees were let go.

When Tenneco purchased Harvester, the Advanced Harvesting Systems idea came to an end. "For whatever reason, [Tenneco[had zero interest in it," Montgomery said. "Absolutely none. The prototype was shipped back to the Burr Ridge Engineering Center, and it sat there for a couple of years and then eventually they just scrapped it out."

Many of the engineers at Harvester at the time were frankly skeptical of the project. Protein harvesting was a high-minded idea that would require expensive research and development, as well as a big shift in the way the entire industry purchased and utilized material.

Ultimately, the times doomed the idea: Harvester wasn't able to stay independent, much less develop such a radical new product.

Rick Tolman left the project in the early 1980s for another opportunity. "They were looking at the future and their present was in doubt, so it didn't make a lot of sense to keep investing in that," he said.

PX-1 Protein Extractor Concept Rendering

▶ This concept rendering by Dean Swanson of the IH Industrial Design group, under the direction of John Hamilton, was the first iteration of a machine for harvesting protein from tobacco plants. It was also part of a presentation to management to get authorization to start the AHS venture group. But because IH management wanted to keep this group as almost a separate company, this concept and any working relationship to IH were severed following the startup.

Ken Updike Collection

Further Evolution

◀ This machine was a later spinoff of the main AHS program, and while it was shown as being used with the Probine design, it was part of a larger system that included a field-going packaging process as well as protein processing. *Gregg Montgomery Collection*

Kelp Harvester

▲ The protein-harvesting system could be adapted to multiple environments—even working in the ocean. *Gregg Montgomery Collection*

Field-Going Packaging

▲ This machine was designed to process and package protein on the fly. *Gregg Montgomery Collection*

THE DAWN OF CASE IH

By Lee Klancher

Derek Stimson remembers the day his father decided to sell tractors like it was yesterday. Stimson's father often would work late at Hi-Way Service, his family's Shell gas station and repair shop in a small town in southern Alberta. Some nights, however, he would come home at five o'clock to spend time with his son.

On a memorable Wednesday afternoon, the father and son were watching television. "I saw an advertisement that showed a Case tractor going up a hill," Stimson recalled. "I thought it looked pretty sharp, and all we had heard about at that time were Massey-Harris and John Deere machines."

Stimson said as much to his dad, who replied, "Yeah, they are good tractors."

Not long after that, Stimson noticed two brand-new Case tractors for sale sitting out front of his father's station. Southern Alberta is blessed with rich soil and big farms; potential customers were driving down the street in front of Hi-Way Service every day.

Simple Beginnings

▶ Derek Stimson, (not pictured), one of the founders of Rocky Mountain Equipment, grew up at his father's service station, Hi-Way Service, in southern Alberta. They started selling Case tractors in 1959 and grew to become one of the largest Case dealerships in North America. *Case IH*

"We sold those tractors right away," Stimson said.

The year was 1959. Hi-Way Service continued to sell tractors in the 1960s, a time that Stimson describes as "tough." Things didn't improve as the decade wore on.

"In the 1970s, it was a terrible struggle," Stimson said. "We had to go to the bank to survive."

When the farm economy turned in the mid-1970s, Hi-Way Service turned as well. Stimson left to get a degree in agricultural technology at The Southern Alberta Institute of Technology in Calgary. He returned home to work at the dealership full-time with his father.

"In 1974, we sold every piece of equipment off the dealership," Stimson said. That same year, Hi-Way Service became the top Case dealership in North America. They did it again in 1976 and in 1979—all out of one location.

When the Case IH merger occurred in 1985, Hi-Way Service suddenly had a much broader range of equipment to sell. The Axial-Flow combines were particularly well-suited for Alberta's wheat fields. Given the fact that Case had discontinued combine production, the red combines were a boon for Stimson's business.

"The merger was a real boost for us," Stimson said. "We went out and sold twenty or thirty combines that first year."

In 1988, a Case IH dealership became available in Lethbridge, Alberta, and Stimson led the purchase of the the dealership. Hi-Way Service became a two-store operation.

A few years later, they purchased the Medicine Hat dealership as well. By 1996, they had a total of seven stores covering all of southern Alberta and the Calgary region. This was good farming country, with big farmers harvesting in volume. Hi-Way Service was moving more than 100 combines a year and continued to add more dealerships into the early 2000s. They were soon moving more than 400 combines per year and had become the largest dealership network in Canada.

In 2007, Hi-Way Service Ltd. merged with Hammer Equipment. A year later, they added Miller Equipment to the fold, forming one of the largest dealer networks in North America. In 2012, the dealerships in the network took on the brand Rocky Mountain Equipment.

Stimson believes the success of his dealerships comes down to very simple principles learned from his dad many years ago.

"I think the key to any successful dealership is service. I have eaten more dinners at farmers' houses than at my own," Stimson said. "You develop a relationship of trust and honesty between you and the customer and it works. It's not a hard thing."

Big Country Dealers

◀ After becoming a Case IH dealership in 1985, Hi-Way Service added dozens of stores and became Rocky Mountain Equipment, one of North America's largest dealer networks. Farmers in their region favored big tractors and powerful pull-behind combines into the 1990s. *Case IH*

THE COLOR RED

In 1985, Tenneco purchased International Harvester's agricultural division and merged it with Case. Steve Tyler was the operations manager at East Moline in 1985, and he recalls those days clearly.

"We'd been through a six-month strike," Tyler said. "Frankly, I can remember getting a paycheck for about three or four Fridays in a row, jumping in the car and running to the bank. Because there was a rumor that they were out of money."

When the news of the purchase hit the plant, the reaction Tyler recalls was positive.

"We were all excited. We were all relieved."

Not long after, the Case IH executives came down to East Moline from Racine. All of the plant gathered to meet them. The only place large enough to hold that many people was the receiving dock. A semi-trailer was brought into the center to serve as a stage, with a half-dozen folding chairs on top of it for the Case and Tenneco executives.

A New Identity

▶ When Case and IH merged, long discussions ensued regarding how to present the new Case IH brand. This logo was one of the proposed graphic treatments. *Case IH*

"Now you've got to remember that at that time, Case tractors that they built in Racine were white," Tyler said. The executives spoke for about an hour, talking about how great it was to be in East Moline and expressing their hopes and expectations for the new company.

"Well, we really didn't want to listen to all that," Tyler continued. "The only thing we cared about was this: What color will we be painting our combines?"

Finally, Tyler recalled, the executives said, "Okay, we've made a decision on the color. The color is going to be red."

"I want to tell you, the crowd just roared," Tyler said. "Even when I think back on it today, after 30 years, I still get goosebumps."

INTERNATIONAL CASE
1660
1020
J. LOWE

Chapter Six

1986–1994 WINDS OF CHANGE

By Lee Klancher

"Farming is a profession of hope."
—Brian Brett, author of *Trauma Farm*

Terry Wolf grew up on a farm near Homer, Illinois. In 1983, Frito-Lay built a receiving plant near his family's farm. The Wolf family sold a lot of corn to Frito-Lay, and Frito-Lay paid for the grain based on quality.

After three or four years of selling corn to Frito-Lay, the quality of their grain was near the bottom of the rankings. The only person they knew that had it worse owned a Massey combine.

The Wolf family had John Deere combines. Terry went to the John Deere factory personally to discuss the issues. Nothing changed their quality.

In 1986, Case IH sent a team to the area to talk with local producers. Terry stopped in to hear them out. "The IH engineers came in and rolled up their sleeves and listened," he said. "They chatted for a couple of hours about what they thought they might be able to do."

In 1987, Terry purchased a Case IH 1680.

Longtime field-test engineer Jim Minihan was doing corn-harvesting research at the time and spent time in the area working with families like the Wolfs.

"We worked with [Minihan] and were able to really get the combine set, learn how it really worked and how to set it best for Frito-Lay and get the best quality we could," Terry said.

The Wolf sample was top shelf, and their earnings from Frito-Lay went up. Minihan became a regular visitor to the Wolf farm, and the family has been running red combines ever since.

While a rotary combine offered plenty of improvements over what existed on the market, the machines weren't perfect. The merger between Case and IH had relatively little effect on the combine development group because Case didn't have combines. For the company as a whole, however, the merger was a tough transition. Plants and dealerships were closed and entire departments let go.

1986 Case IH

◀ While the Case IH merger had a traumatic effect on the company, the combine line came through with very little change. Case didn't have a combine line, so the IH equipment was simply rebranded and the team building them remained largely unchanged. The early 1600 Series machines wore the old Case IH lettering with "Case" much larger than "International." *Wisconsin Historical Society #117312*

Case IH Design Sketch

▲ After the purchase of International Harvester by Tenneco in 1985, the revised series was sketched with new Case IH badging. This 1680 wears a Case IH logo that was not used. *Gregg Montgomery Collection*

The fact that the farm economy remained flat horrid didn't improve matters.

When the merger hit, equipment dealers Ed and Gerald Heim had been selling red combines since 1955. Gerald recalled how hard it was. "When the merger came out, and, of course, right before that, it was really chaos," he said. "I think 21 banks went bankrupt or broke out here within 100 miles of us, and my bank went down. International Harvester went down. Heston Corporation went down. Orthman manufacturing went down . . . most all of them companies just went broke, so there we sat with no suppliers and no bank. Then, of course, the repossessions all started showing up.

"I tell you what, that was a super tough time. We don't want to see those again."

The merger helped things, as the newly formed Case IH had more cash and was able to improve parts availability. A new bank opened up in town, and Heim was able to get financing.

He also recalls that the 1600 combines' reliability took a dip. Straw spreaders and skid shoes under the header didn't last. Roller chains were cheap. Engines in the early machines showed up with 20 to 30 percent less power than they were rated to produce. Welds on the rear axles broke.

Jamie and Steve Horst, who were part of the management team at the East Moline plant for many years, remember the entire plant mobilizing to repair the 1680 rear-axle welds. In fact, much of the work was done in tents.

"I think we had something like 300, 500 combines in the yard," Jamie said. "That's when we had those circus tents set up out in the parking lot."

"[We had] three shifts of repair people out there repairing combines," Steve added. "We had a major issue, but we addressed it head-on, got it, set up, ran it, finished it—done."

Ed's father, Gerald Heim, was on the Combine Product Advisory Board at that time, which meant that communicating issues was something he was expected to do. And he was vocal about his issues. The key to being able to make it through all that, he said, was the people who listened. He worked closely with Harvester's team, including Camiel Beert, Gerry Salzman, and Jim Minihan.

"Those guys just gave everything they had, way over and above the call of duty over the years," Gerald said. "They'd drive all night just to take a custom cutter apart, or they work on his machine out there in the dark just to help keep it running, to help support the Pro-Harvest program, and so they're very, very dedicated people."

The work to improve reliability was tackled from two ends. Dr. Glenn Kahle, who led the engineering team from 1979 to the early 1990s, remembers implementing failure-mode analysis to the combine line in which each part was analyzed statistically to predict when it would fail.

Prior to that program, he said, the attitude was "if something broke, fix it. There was no real anticipation of how long it was going to be before something else broke."

Kahle added, "We introduced the same reliability program that we had on the Magnum tractor. That took care of a lot of things."

1500 Prototype

◀ This prototype wears a "1500" moniker.

Gregg Montgomery Collection

Ed Heim may have been frustrated with those little things, but the performance of the rotary combine at that time kept him and his family firmly in the fold. The increased horsepower and an improved cleaning system on the 1600 Series machines were just what they needed to sell in their part of the country.

"That became the time that we really turned up our demonstration program out here against John Deere, which was our major competitor," Heim said. "We went door to door during fall harvest. We just went field to field and pulled in and showed people what we could do. It was amazing how many people realized how much better a job we could do. We had times we took a tractor with a grain cart that had scales on it. We took that with us, and we would harvest the same number of rows of corn and show them how many more bushels per acre they could get just by running an Axial-Flow combine. Of course, the grain quality was always better.

"It was such a difference that we were able to trade for a lot of competitive machines. The word in our area was, basically, if you were a progressive producer, that was the combine to have."

Case IH 1620

▶ The 1600 Series was introduced in 1986 and featured increased grain-tank capacity, larger concaves, and more separating area than the 1400 Series combines. This Case IH 1620 is owned by Josh Olson. *Lee Klancher*

Case IH 1640

▶ The 1600 Series combines were introduced with International Harvester engines. In 1989, the series switched over to CDC engines in models designed as XPE editions. The change was not visible externally. This 1640 is owned by Randy Stokosa. *Lee Klancher*

Case IH 1660

▲ The 1660 had a DT-466B IH engine early on, and a CDC engine later in production. This one is owned by Adam Tordai. *Lee Klancher*

Case IH 1660

▲ The IH-engineered 1660 had 10 more horsepower than the 1460. *Lee Klancher*

Case IH 1680

◀ The largest of the 1600 Series line was the 1680, which had a 225-hp DTI-466C engine. Many of the 1680s with tracks were used for rice harvesting, with a good percentage of those destined for work in California. *Wisconsin Historical Society*

Case IH 1680 Rice Special

▶ The 1680 was available in a rice edition with the larger tires and other upgrades suitable for harvesting rice. The key to the machine's rice capabilities was a special rotor for rice. *Case IH*

Case IH 1682

▶ The big pull-behind model was also upgraded to the 1600 Series. It was built in 1987, then from 1990 to 1991. *Case IH*

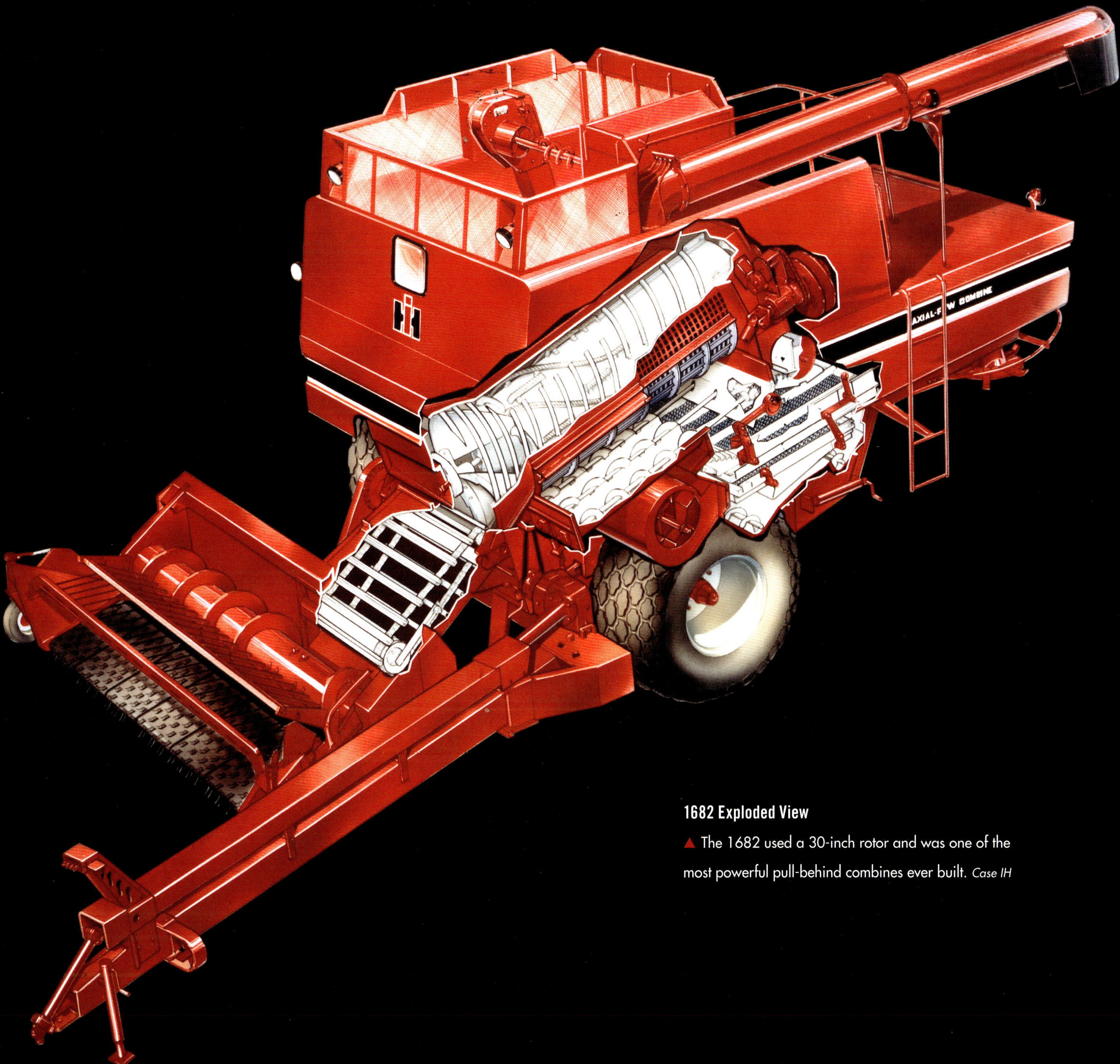

1682 Exploded View

▲ The 1682 used a 30-inch rotor and was one of the most powerful pull-behind combines ever built. *Case IH*

Case IH 1670 Hillside

▲ The Hillside model was also upgraded to the 1600 Series and produced from 1989 to 1992. *Case IH*

Case IH 1666

▲ Second-generation 1600 Series models had only minor upgrades from the first generation. The series had a new cross-flow cleaning fan and an improved cleaning system. The 1666 and 1688 had larger air cleaners, as well. *Case IH*

Case IH 1644

▲ The second-generation 1600 Series was produced in 1993 and 1994 and included the 1644, 1666, and 1688. *Ken Updike Collection*

Case IH 1688

▲ In 1994, the second-generation 1600 Series combines were upgraded with a straw-chaff spreader, radial-seal air filters, and an optional cold-start package with engine-block heater and ether injection. This one is owned by Darius Harms. *Lee Klancher*

GOING PUBLIC

In the mid-1990s, when Case IH went public, those close to the company said that it had a tremendously positive effect on both morale and day-to-day operations. For one thing, a publicly-traded company had to be profitable, and that hadn't been the case under Tenneco, whose oil and gas holdings were generating tremendous profits—the company didn't need Case IH to turn a profit.

Tenneco's deep pockets funded critical programs like the development of the Magnum tractor, an expensive endeavor that was vital to the company's long-term survival. With a struggling farm economy and a historic debt load, Case IH was bleeding cash. In the late 1980s, the subsidiary was losing hundreds of millions of dollars each year.

The Magnum tractor was the right machine for the market and it sold well in the late 1980s and early 1990s, as did the Axial-Flow, for that matter. Sales of the machines improved matters, but that wasn't enough to balance the books.

Restructuring in the late 1980s was supposed to turn things around. In January 1989, Tenneco CEO James L. Ketelson told the *Chicago Tribune* that he expected $100 million in profit soon, and an increase to $400 million in the early 1990s.

He was wrong. The farm economy tanked, and Case IH posted a $1.1 billion loss in 1991.

By that time, Tenneco was publicly announcing its intentions to sell Case IH, but finding a buyer for a heavy-equipment manufacturer posting billion-dollar annual losses was a tough row to hoe. Case IH executive Dana Mead later admitted that the sale price dropped as low as one dollar.

No takers.

Case IH determined the only road to salvation was to put renewed effort into finding a path to profitably.

Robert J. Carlson took over the CEO position at Case IH in July 1991. Carlson had been an executive at Deere & Company for nearly 30 years, and expectations were high that "Tractor Bob," as he was known, would put the company back on track.

Carlson was a likable leader who is remembered fondly, but he wasn't able to turn the tide.

Michael Walsh came to Tenneco in 1991. He had built his career at Cummins and Union Pacific Railroad and focused much of his efforts on working with new Case CEO Edward J. Campbell to implement tough cost-cutting measures and new structuring. But Walsh was felled by a brain tumor that would eventually take his life.

Jean-Pierre Rosso came to Case IH in 1994. The French-born Wharton School MBA graduate had a long career with Honeywell and had also served as the CEO of Rossignol Ski Company.

He was lured away to help engineer a turnaround for Case IH. "I was approached with a marvelous opportunity," he said in a *Racine Journal Times* article. "It was obvious this was and is an enormous opportunity for success."

A key to his recruitment was the $1.4 billion that Tenneco had alloted to help with the restructuring. The tough cuts made by Walsh and Campbell meant Case IH was poised for growth.

"This is going to come to the top of the list as far as successful turnaround stories in America," Rosso said.

Rosso came on-board in April 1994, as the CEO of the newly formed Case Corp., which owned Case IH.

Case Corp. was formed when Tenneco sold 25 percent of Case IH shares to the public. The independent company formed as one of the 150 largest independent firms in America.

In the first quarter of 1994, the company posted earnings of $81 million, a terrific turnaround that allowed Tenneco to go public, selling 56 percent of the company as shares. The farm economy surged, and sales of machinery grew dramatically.

Rosso proved a smart, capable leader. Profitability continued to increase, and Case IH would eventually stand on its own—without any Tenneco ownership.

PROHARVEST SUPPORT PROGRAM

By Ron R. Schmitt

"It is our belief that every customer is entitled to and shall receive two distinct services . . . one from the product itself and the other from the organization back of it."
—Cyrus McCormick, *The McCormick Promise, (1878)*

When the International Harvester Company was formed in 1902, a key operating principle that carried over from the McCormick company was to place as much emphasis on customer service as it did on product development. IH experienced strong sales of the Axial-Flow models during the first year of production in 1978, and made significant gains in market share. Company executives would need to apply the McCormick Promise to ensure the continued success of the Axial-Flow combine.

Dan Kennedy, manager of Product Support for the Midwest region, explained how the program began. "It became clear early on that we needed to initiate some critical customer and dealer support action as few in the field were familiar with the new rotary technology." He was asked to seek input from others and structure a program. Many questions were coming in from customers and dealers in the field regarding setting, troubleshooting, and operating the new combines in less-than-ideal field conditions.

Field Support

▶ A ProHarvest representative meets with the owner of a new 9230 Axial-Flow combine to check on the machine's performance and assist with settings. *Case IH*

"Some IH users were apprehensive, and the competition became very aggressive in their focused effort to discredit the Axial-Flow rotary principle," Kennedy said. International Harvester had much riding on the Axial-Flow line, with millions of engineering hours and dollars invested.

The implementation of the program that later became known as ProHarvest Support (PHS) started small in 1979 with several training instructors from the Kansas City Midwest regional office being selected along with local territory service managers (TSMs) from the harvest areas. According to Kennedy, "The new PHS field team received training and spent time at the East Moline Works with Camiel 'Cam' Beert, Bob Francis, and several other Axial-Flow experts, and they spent time with Jim Minihan in Product Reliability at the Product Support Center."

Much of the expertise on the Axial-Flow combines resided at the East Moline plant, so the transfer of knowledge to the ProHarvest field team members was vital to the success of the program. Team members were then sent out to help customers and dealers. The budget was very tight, so the team used Dodge K-Cars and IH Scouts with magnetic "ProHarvest Support" signs stuck them on the doors.

A small team was tasked with further development of the PHS program in 1981. Key participants included Gary Wells (product development); Vaughn Allen (product support); Gerry Salzman (product marketing); Dan Kennedy (service marketing); Jim McLain and Elliot Bourgeois (corporate parts); and Ed Powell (TSM, Kansas City). This team refined key objectives, execution plans, and funding requirements.

Early objectives were teach operators how to set Axial-Flow combines in varying conditions, help dealers with technical problems and questions, and feed information back to the Product Support Center and factory to make improvements. The PHS team was to move along with the harvest and help resolve problems.

Combine Patrol 1964

◀ Ed Powell on "Combine Patrol" in 1964 with an IH service truck. Powell's efforts supported the 03 Series combines and canvassed combine users from Texas to Canada. *Ed Powell Collection*

Over the next several years additional TSM personnel were tapped to focus on PHS only as the harvest moved through their areas. One of the early things that the PHS team learned from the field was that the initial machine settings in the operator's manual were not the best and in some cases did not work in tough conditions. Published rotor speeds and concave settings were too slow and tight. The team soon had new recommendations available for customers and dealers. It also was soon discovered that Axial-Flow combines could harvest more crops besides wheat, corn, and soybeans better than conventional machines, with one particular crop being fescue grass. Recommended machine settings were soon developed for each added crop.

Custom cutters started purchasing and switching their fleets over to Axial-Flow combines in the early 1980s. Only a few custom cutters were utilizing IH combines prior to this, and the PHS team quickly reached out to these operators. As the number of custom cutters operating Axial-Flow combines grew, dealers often became overwhelmed when the custom cutters moved through their respective areas. PHS personnel realized the need to provide more support to the cutters and the local dealers along the harvest paths. Typical operators would have six or more machines, and there could be as many as 25 operators in a given area at a time. A couple of Ford vans were approved and added to the PHS vehicle fleet in 1983. They were decaled and loaded with support tools and materials that really helped the PHS program take off.

PROHARVEST SUPPORT EVOLVES

Ed Powell had been providing periodic assistance to PHS since 1980 due to his experience working with combines back in 1964 during the "Combine Patrol" program. After retiring from IH, he returned as a consultant and was fully involved in PHS in 1984. Powell and his wife, Vonda, took their motorhome out on the road each year, moving with the harvest from Texas to Montana. In addition, a specially outfitted semi-trailer was added with parts bins, a monorail crane, and a computerized parts management system that was cutting edge at the time. Inside the trailer was a select inventory of over 3,000 part line items, some of which were fast-moving items for dealer backup, such as belts and hoses. Many others were harder to find parts, including two spare engines, rotor gear boxes, transmission drive components, hydraulic pumps and valves, bearings, sickle-drive boxes, and even sheet metal parts for headers. All parts sales were billed to and through the local dealers—nothing was sold directly to customers. As parts were consumed from the trailers, the parts depots provided replacements via next-day delivery.

Dealer Badge

▶ In 1992, dealers who qualified received this badge of honor. *Case IH*

Les Hearn, who had just retired from IH Product Support, joined the PHS program in 1985. Powell stated that he and Les (and their wives) parked their motorhomes at the dealerships with the PHS parts trailers for the first few years for visibility and availability for late-night parts pickups. The first years also involved much fact-finding for the team as it worked closely with various dealerships along the harvest path, quickly learning their parts and service capabilities.

Under the leadership of Powell, Hearn, and Kennedy, PHS evolved to be more focused on the custom cutter operations as the objectives were refined to "help enhance dealer support when custom harvesting crews came through an area and provide backup support to the dealers . . . but always lend a hand to any local customer or dealer," Kennedy stated. Powell said the PHS team also added a new objective "to not have any machine down more than two days, either for parts or service. That was a stretch at times, often requiring overnight travel to obtain parts."

Custom cutters put thousands of separator hours on their combines during the harvest season, running seven days a week for up to six months in duration. They are the lead machines out in the field, so if a problem develops on their combines, it eventually will show up elsewhere on other customers' machines. The team realized that design and manufacturing quality problems could be quickly detected and resolved by working with custom cutters. This allowed an opportunity to develop plans to address those problems on other customers' machines.

Ed Powell learned from his combine patrol experience that "custom cutter crews are family—no matter what color combine they run. They shared information and helped each other." The custom cutters know what works and what does not, and they provide invaluable feedback to engineers for future improvements.

Executive management also knew that competing manufacturers did not have similar programs at the time. The high-quality support could help convert non-IH/Case-IH custom cutters over to Axial-Flow combines.

The PHS program became stronger after the Case-IH merger, and in 1986 was further expanded to add personnel and a second semi-trailer for parts, as two parallel routes (east and west) were established from Texas to Montana to provide better coverage across the harvest area.

Additionally, an annual two-day PHS kickoff and open house in Vernon, Texas, was held in mid-May. All the custom cutters in the area were invited to the event, no matter what color combine they ran. They also asked the cutters their opinions and what Case IH could do to earn their business. A product meeting was held to discuss the new Axial-Flow combines and update the operators on planned PHS activities. Machine maintenance and setting information was discussed and important safety training was stressed.

That first year, approximately 40 to 50 people attended the open house. Several years later, annual attendance had increased almost tenfold.

The ProHarvest Qualified Dealer program certified dealers with adequate facilities to service Axial-Flow combines. Dealers that met all the requirements received a "PHS Qualified Dealer" decal to place on their main entrances, plus a designation in the booklets given to custom cutters.

Favorite Stop

◀ One ProHarvest Support stop along the harvest path in July 1984 was Greensburg Equipment in Greensburg, Kansas. Field representatives used the dealership as a base of operations. Note the company Chrysler K-Car in foreground. The dealership building, and much of Greensburg, was destroyed by an EF5 tornado in May 2007. *Ed Powell Collection*

ProHarvest Support was dynamic, and it took about five years before it was fully understood by dealers and customers such that all parties enjoyed its maximum benefits. Long days and quick decision-making were critical.

"Every day was twelve to eighteen hours, some overnight, seven days a week from May to October," Powell said. Often, the PHS representatives worked alongside the dealer and custom cutter mechanics, consulting, troubleshooting, and sometimes assisting with repairs to get the machines back in the field as quickly as possible.

"If you had to make a decision on something, it had to be within thirty minutes to an hour," Powell said. Sometimes if a machine required major repairs it might be sent ahead of the harvest to a dealership that had the capability to make the necessary repairs. Communication was critical. "We were in constant contact with custom cutters twenty-four hours a day," Powell said. "It wasn't uncommon for us to have over one hundred phone calls a day.

"If we had a customer that was having problems with his machine, we were also in contact with his dealer at home, which might have been in Minnesota or North Dakota or Canada. We also worked with them on what they wanted us to do and we would make it happen. In some cases, they would fly their own people to where the problem was or they would

Let 'em Run

▶ Pro-Harvest Support representative Ed Powell watches Axial-Flow combines run during harvest.

Ed Powell Collection

send a truck with whatever was needed . . . there were three things to consider whenever you had a situation: the customer, the dealer, and the company. And all must be treated fairly and equitably."

Being resourceful was also a key requirement for PHS personnel. As Powell recalled, "One time we had to change a hydrostatic drive pump at midnight. There was no loader or lift available in the field, so the unloading auger was used as a boom along with a winch to remove and install the pump." Another time, he remembered, "We had a two-speed Hydro motor blowup on a Saturday, the first year they came out. No new motor was available, but there was a mating pump available on one of the parts trailers. I looked at the parts book and noticed that some of the components were common between the pump and motor. . . . We pulled the old pump and motor off, cleaned the hydro lines, and installed the new pump. The old pump and motor were disassembled and we salvaged enough compatible parts out of both to rebuild the motor and fix it. It ran that way for at least three years."

One of the more notable problems the PHS team dealt with was the Consolidated Diesel Company (Cummins) engine failures, which began showing up early in the harvest. Engineering was quickly engaged to come up with a fix. "We finally got dealers lined up to where we overhauled at least one engine per night," Powell said. "If a customer had a machine in by five p.m., we would give it back by five p.m. the next day." The engine overhaul was major and included new pistons, rings, sleeves, and bearings, as well as pulling the engine off the machine and reinstalling it.

Another serious problem included radiators getting plugged with chaff and overheating the engines daily due to a new core design. The PHS team led the change back to a conventional core design. "We had many situations, but our dealers always came through for us, and we had tremendous cooperation throughout the company," Powell stated.

The PHS program had good support from Case IH management through the transition years, but several PHS team members thought it would be good if senior executives in Racine better understood program. Powell suggested and Kennedy (now manager of Customer and Dealer Support in Racine) proposed the idea of holding a field day for select senior management, including then CEO James Ashford. Powell then contacted several custom cutter owners and invited them to the event for some "one-on-one" discussions with Case-IH management.

The first field day, held in Onida, South Dakota, was so successful that a second field day was organized the next year in Glasgow, Montana. Both years operators had machines down, resulting in a critical need for parts that were flown in on the company plane. On the Glasgow trip, two executives were bumped and seats removed so the needed parts could be loaded in their place at 4:30 a.m. and accompany CEO Bob Carlson to Montana.

These situations received a lot of attention throughout the company, and more importantly the custom cutter community—the CEO was delivering parts for ProHarvest! The management team arrived early that day in Glasgow for a working breakfast followed by one-on-one discussions with operators throughout the morning and a working lunch, before going into the field where four combines and about 100 acres of wheat awaited them.

One new initiative that came about from the management field days was the Field-Ready program. During the Glasgow visit, Powell told one of the vice presidents during lunch that one of the biggest constant complaints he heard from the dealers along the harvest path was that the new combines should be field-ready when they leave the factory. Dealers were spending up to a week getting a machine ready for delivery, fixing numerous problems like misaligned chains, sprockets, belts, and pulleys, loose bolts, unpainted areas, and numerous other adjustments necessary to make the machines run and look like they should. Nothing more was discussed on until the following December when Powell received a phone call from the vice president, who asked him to come to the plant for a month and audit the combines for the custom cutters (which were built at the beginning of each year's production run). Bill Durke and Powell spent a month at East Moline working with quality inspectors and the Manufacturing Operations teams, pointing out the various problems dealers were finding. After the first year, warranty costs to the custom operators were down significantly, enhancing combine product line profitability.

The ProHarvest Support program is often overlooked in the annals of IH history. But it was a key element to the success of the Axial-Flow combine. It brought focus to needed product support and made a great product even better. ProHarvest accomplished its initial objectives and proved to be very successful and beneficial for International Harvester and Case-IH over the years. The program has evolved and grown significantly since its inception in 1979.

The Fleet

▲ The new ProHarvest Support vehicle fleet stands ready to follow the harvest from Texas to Montana, assisting Axial-Flow combine operators. *Case IH*

The ProHarvest Support program continues within the Case New Holland/Case IH enterprise under the leadership of David North. He took over the well-established program in 2010 and is laying the groundwork for the future. Key program objectives of supplying parts, technical advice, and operational instruction to Axial-Flow operators remain the same.

Thanks to the determination, dedication, and support of both IH and Case-IH personnel at many levels, from the early development to today's modernized program, the McCormick Promise continues. The ProHarvest Support program has been and will continue to be a key component that established IH and Case-IH as a leader in combines.

Chapter Seven

1995–2008 THE ART OF NEW MACHINES

By Lee Klancher

"The scientist describes what is; the engineer creates what never was."

—Theodore von Kármán, celebrated twentieth-century physicist and aeronautical engineer

Jim Lucas started working for International Harvester (IH) in 1967 as a test engineer in the combine product development group at the East Moline, Illinois, plant. He developed combines for many years, and one of his projects in the early 1980s was engineering the popular 1000 Series grain headers.

"We got the grain headers and corn headers all in production," Lucas told his boss at the time. "Well, if I don't have anything more to do next year, I'm going to spend a lot of time on the golf course."

The odds of his boss agreeing to this proposition were extraordinarily low. Lucas was one of IH's most talented engineers, and the company was in desperate need of new products at that time. Rather than being sent out to golf, Lucas moved to a new role as design leader of the chassis group developing the new Magnum tractor. He created the master engineering plan and managed the supporting component engineering teams to meet the project timelines and design requirements. The resulting Magnum tractor set new industry standards and kept the legacy of red agricultural machinery alive.

Unfortunately, while the company focus was on developing the new Magnum, the combine group lagged behind with limited budget and attention.

In the mid-1990s, the combine line was in need of further evolution and capacity increase. The styling hadn't been changed drastically since the introduction of the line, and the market demanded a larger, more powerful machine.

The first step taken by the company was to freshen up the existing line of combines with a new cab, more efficient hydraulic system, and increased power levels along with a host of minor improvements. The result was the 2100 Series that launched for the 1995 harvest season.

The striking new cab prompted many to wonder why the rest of the machine hadn't been restyled as well. A lack of

Good Times

◀ Starting in the mid-1990s, the situation on the farm improved. Crop prices rebounded, and farmers who survived the hard times of the late 1980s and early 1990s were able to grow. For the farmer to continue to thrive, bigger combines were vital. *Case IH* budget is the simple answer. The late 1980s and early 1990s had been brutal times for the farmer, which meant sales were low for all the agricultural equipment manufacturers. Budgets were tight, and the research and development done on red combines was minimal at best.

This was about to change.

Leadership at Case Corporation was transitioning, with Michael Walsh brought in to lead Tenneco, the parent company. Walsh pushed to cut costs and increase investment in appropriate areas. He brought in Dana Mead, an MIT graduate and U.S. Army veteran with a strong record of business and military leadership. Mead in turn hired Steve Lamb and Jon Carlson, a pair that would change the direction for red combines.

Carlson was the son of "Tractor" Bob Carlson and a former snowmobile racer. "My philosophy is to go for it because life is short," Carlson said in a profile in *Irrigation & Green Industry* magazine. "Go fast every day. If I had to do it over again, I'd go faster and take even more risks."

CBX Combine Design Sketch

▼ This early sketch shows the CBX combine being designed by Case IH in the late 1990s.

Case IH

Like his father, Carlson spent a good portion of his career working for the green brand. He came to Case IH in the mid-1990s and spiced up the management with a helpful dose of energy and ambition.

Steven Lamb became the chief operating officer at the Case Corporation in 1995 as Tenneco spun off the J. I. Case Company to become Case Corporation, a publicly traded company. A West Point graduate with a Harvard MBA, Lamb became a U.S. Army company commander before moving on to a successful career as a business executive. He worked closely with the president of Tenneco in the late 1990s. Lamb's work at Tenneco helping Case to develop more efficient systems led him to join Case and direct its European operations in 1993.

Lamb and Carlson worked closely together in the mid-1990s. With the farm economy rising and the red combine line showing its age, the pair spearheaded an aggressive strategy to take back some market share with a combine more significant than the "lipstick on a pig" approach of the revised cab on the 2100 Series.

In the mid-1990s, the largest machine in the Case IH line was the a Class 6 combine. This was the most popular size at the time, but that would change in the coming years. Case IH management understood the market was trending toward larger combines, and executives pushed to create an all-new machine that filled that gap and, in fact, went well beyond that.

The new machine—dubbed the CBX—was to be powered by a 600-horsepower engine and could carry a 1,000-bushel grain tank. "They figured we could kick Deere's butt with one mega combine," Gerry Salzman said.

The program received solid backing and the kind of R&D budget the red engineering team hadn't seen in more than a decade. This program was what lured Jim Lucas back to the combine fold.

"This was a once-in-a-lifetime opportunity," Lucas said. "We were going to develop a new combine from the ground up."

CBX Scale Model

◀ This scale model shows the CBX combine design, which was a radical departure at the time and holds up well even today. *Case IH*

The design team was given a clean sheet to build the CBX. He couldn't pass up that opportunity.

The CBX combine required a significant commitment on many fronts. For starters, the new machine was too big and too tall for the East Moline Plant to build, so a new assembly facility was needed.

Jesse Orsborn, the engineering program manager on the project, said the new CBX was intended to be the highest capacity combine on the market. The team was aware the others were developing large combines. The CBX was intended to be bigger than anything anyone else was even considering.

"We knew at the time that the New Holland engineers in Belgium were developing high-capacity combines as well," said Don Watt, vehicle systems engineering manager at the time. The New Holland development machine was designed to fit into Class 6, 7, and 8—big, but not as high-capacity and powerful as the CBX.

The engineering group was also aware that Deere & Co. was getting serious about bringing a rotary combine to market so the time was ripe for a new Case IH combine.

The group needed a new facility to house the development team for this new combine. Management held a strong desire to move the new development team offsite from the East Moline combine plant to allow it the laser-like focus needed to develop such an important new addition to the family. A vacant construction equipment plant in Mount Joy, Iowa, turned out to be the perfect fit. The office could seat the large development team, and it was adjacent to a well-equipped multi-bay shop large enough to develop this new giant combine. This allowed the group to put everyone needed to create the CBX under one roof and to build the early prototypes right next door.

The CBX development team consisted of not only the traditional design and product evaluation engineers, but representatives from every area of the company who would be impacted by the development and eventual introduction of this new combine.

The team would be able to work in person and quickly, and it included engineers who analyzed designs and created bills of materials, on-site driveline and hydraulic system engineers, and reliability engineers who would lead the product reliability planning and monitoring as the design proceeded.

They also had people in charge of purchasing prototype parts and developing long-term supplier agreements for those parts, technical writers to develop support documentation on site, and administrators who monitored budgets and timelines.

The team was challenged to develop not only a new product, but a new product development process that was envisioned to be used for all future Case Corporation product development programs. The organization and management philosophy was patterned after the very successful Platform Engineering model used by Chrysler in the development of its highly successful minivan line.

Given a clean sheet, the engineers developed a variety of innovative features. The CBX was designed with:

- A large 36-inch diameter rotor.
- An all-new variable-displacement hydrostatic drive that had wheel motors on all four corners.
- A novel, simple all-new cleaning system designed to minimize the effects of side slopes.
- An innovative new grain handling system.
- High-capacity vertical straw and chaff spreaders with greatly improved residue management.
- Belts and chains were eliminated from the entire machine by the use of hydro-mechanical rotor and feeder drives paired with gearboxes that replaced the previous belt and chain drives.

"One of the key requirements of the CBX combine was to not only develop bigger and higher capacity combines, but to build in much higher reliability than previous combine models," Lucas said.

One of the young engineers added to the development team was Jay Schroeder, who was assigned to develop several of the gearboxes that would replace the belts and chains used on the combine's current architecture. Schroeder grew up on a dairy farm near Coal Valley, Illinois, where his mechanical aptitude quickly made him the family go-to guy for repairing broken machinery.

One of his first regular tasks was fixing chain links on an old Gehl forage wagon and performing maintenance and repairs on the pull-type forage harvester and mower conditioner. "It was usually something down deep in the machine that you had to try to figure out how to fix," Schroeder said. "You could typically hardly reach what you needed to repair no matter whether you were on top, underneath, or beside it, and it made me think, 'Well, why did they design it like that and how did they do that?'"

"I remember countless times my dad would say, 'I wish I knew the engineer that did whatever, whatever, whatever,'" Schroeder said. "Those things just kind of stick with you as you grow up, and you start to think as you get older, 'Yeah, why did they design it like that?'"

That led him to engineering school, which led him to a job working as a drivetrain engineer for Case IH at the Tech Center in Burr Ridge, Illinois. After several years in drivetrain engineering at Burr Ridge, Schroeder was chosen to work with Jim Lucas and Pat Dinnon and charged with developing the new hydro-mechanical rotor drive for the CBX. When he asked why the rotor drive was so critical, he was taken out back to the engineering shop and shown a (then current) combine.

"Okay, now you have this combine that's been running at peak power all day; it's full of dirt and dust, and now your rotor

CBX Prototype

◀ Working prototypes of the CBX were in the fields by the late 1990s. *Case IH*

belt breaks!" Schroeder was told. "You have to crawl into this compartment, you're out in the middle of the field, you have this hot engine and material around you, dust and dirt, and now you have to maneuver this big belt into the pulleys and sometimes you do this two or three times a day." Schroeder recalled his time changing chains on the farm and immediately saw the light. This was his chance to fix one of those things that had made him ask, "Why?" when he was a kid.

For the entire engineering team, the CBX was a chance to simplify and make more efficient the original Axial-Flow combine. The chains and belts would be eliminated, key components made more reliable, and the engineers believed they could double the reliability of the combine in the hands of the customer.

"That's a quantum leap," Schroeder said. "I was thrilled to be part of it."

The project management team had great enthusiasm for the CBX project. For the experienced engineers on the team, it was a wonderful opportunity to utilize their experience for more reliable, higher capacity, and simpler designs. The functional systems engineering (everything that the crop touched) was the responsibility of a small handful of experienced engineers led by Rich McMillen, Orlin Johnson, and Bob Matousek. Jim Lucas led the systems engineering team to develop the engine systems, drivelines, hydraulics, cab and electronics, and the structure. Many recent engineering grads were hired to fill out the ranks. The North American agricultural and construction businesses were in the doldrums, so Case was fortunate to be able to hire the best and brightest graduates from the top Ag Engineering programs in the United States. For the young engineers coming to the project, it was a marvelous, once-in-a-lifetime opportunity to work with some of the finest combine minds in the business. The team had several years to work on the concept, and it had working prototypes in North America as well as Europe that showed great promise.

Promise, sometimes, isn't enough.

2100 SERIES

By Ken Updike

Three models were offered in the new 2100 Series. The 2144 rated at 180 hp with a 145-bushel grain tank; the 2166 rated at 215 hp with a 180-bushel grain tank; and the 2188 rated at 260 hp with a 210-bushel grain tank. The 2100 Series was introduced in December 1994 and started production in 1995.

Case IH marketed the new 2100 Series with slogans such as, "You've never seen a harvest like this," and "Everything else is history!" The major new component in the 2100 Series was an all-new cab. In addition, all-new body styling featured rounded corners, new decals, new shielding, and a new wider ladder.

Case IH 2144

▼ This 1996 Case IH 2144 was photographed in Dover, Minnesota, on Wegman Farms. The 2144 is the smallest Axial-Flow model in the 2100 Series. *Rachel Wegman*

The all-new cab sported curved front windshield glass and repositioned forward cab, posts for better visibility when using larger 20- and 30-foot headers. The windshield was 23 percent larger (8 inches wider) than the 1600 Series' glass. Behind the cab, a window was added to the grain tank to allow the operator to view it from the seat. This new cab was easily identified by its curved red plastic rooftop.

The full-length door/window on the cab's right-hand side could be opened for communication, service access, and cross ventilation. It could also be used as an emergency exit. A fixed ladder was located in front of the right-hand service deck.

The cab used four rubber iso-mounts precisely aimed toward a point just above the operator's head. This created a "virtual cab suspension" to minimize vibration. The design effectively aligned both the cab's and the operator's centers of gravity.

Inside was a new control and instrument layout. The previous L-shaped handle that controlled the hydrostatic drive and direction on the combine was replaced by a joystick-type handle with multiple push buttons to control the combine's most-used functions. These push-button switches controlled the header height, reel fore-aft, unloader auger in-out, and unloader auger on-off. Access to these buttons with the thumb meant the operator no longer needed to reach for them, as was the case in previous models. The cab also had two interior dome lights and a console light, versus the previous cab's single dome light.

Combines fitted with the Field Tracker had a head that could tilt laterally, a function also controlled manually from the joystick handle. (This multi-control propulsion handle is still used in today's Axial-Flow combines; however, it has been greatly improved with more control features.) To the right of the handle in the 2100 Series was the all-new control console. Here, with the flip of a switch, the operator could control reel speed, separator engaging and disengaging, header height and sensitivity, rotor speed, cleaning-fan speed, and concaves. A flip up of the armrest allowed the control rates for raising and lowering the header to be precisely tuned to the desired speed. The entire console moved with the seat to keep all controls where they were needed and expected.

Beside the new console was the range transmission shifter lever. This was moved from its former location directly to the left of the operators' seat. By moving the lever, a training seat (a.k.a. "buddy seat") could be added. The cab also featured a new operator's seat with improved suspension.

With the key on, a new cab blower/pressurizer located under the buddy seat ran constantly to provide air movement inside the cab. A new cab air-filtering system had an improved filtration system with up to four times more capacity for longer service intervals. In addition, the cab air filter was relocated for a cleaner air intake port.

The new A-post instrumentation included a multiscreen digital data bank. Analog (needle) gauges were still used to read the fuel tank level, engine coolant temperature, and electrical system voltage level. These plug-in style gauges were the same as those used on the original 1400 Series combines.

The new cab was pressurized, meaning that when the cab power solenoid was activated, the blower began operating. Once the operator set the desired cab temperature from 65° F to 85° F, an auto temp control took over. Using three solar sensors to measure cab temperature, this feature activated the AC or heater to maintain the desired temperature. (The sensors were located in front of the steering column base, at the left-hand side of buddy seat, and at the right-hand side of the lower fuse and relay bank.) This was a handy feature because inside cab temperature can vary with the direction of travel—if the machine is driving into the sun, the cab will heat up much faster than if it is traveling away from the sun or in the shade.

One design feature of the new cab was that the cab posts served not only as structural members, but as the cab's air ducts. Many operators drilled holes in these posts to mount radios or other monitors. One owner who did just this complained to his dealer that the new cab whistled. When an infield service call was made, the owner's mistake was pointed out. Luckily, some sealant stopped the whistle.

A new entry ladder that used wide cast-aluminum steps and large tubing handrails was a distinct feature on the 2100 Series. This replaced the narrow, stamped-steel ladder that had been used since 1977. (The cast steps were the same steps used on the 71-7200 Series Magnum tractors.) An optional sixth step held on with two rubber strips could be added to ease access to the ladder. The strips allowed the step to deflect

Case IH 2166

▼ The Case IH 2166 is the midrange combine in the 2100 Series. It is also the most popular model, outselling the 2144 and 2188. *Case IH*

if it hit an obstruction. The ladder could be pivoted backward or rotated forward if needed. The forward setting was recommended during transport. When rotated forward, plastic safety chains blocked the ladder opening so the operator could not walk off the platform inadvertently.

The use of precision farming technology was expanding rapidly in the 1990s. The 2100 Series Axial-Flows were the industry's first combines to be factory-fitted with precision farming technology equipment. Case IH termed its precision farming AFS (Advanced Farming Systems). The acronym is still used today. An easy indicator of a 2100 Series Axial-Flow factory-fitted with AFS is an elliptical-shaped AFS decal on the combine model stripe. Another giveaway was a round white plastic AFS receiver dish at the front edge of the cab roof, centered above the wiper.

Case IH 2188

▼ The 2188 is the largest model in the 2100 Series. This 2188 is equipped with a folding grain bin extension for extra grain carrying capacity. Owned by Andrew Tucker. *Lee Klancher*

Underneath the newly styled body was the traditional Axial-Flow rotor and cleaning design. The new 2100 Series offered a choice of two factory-installed rotors—standard or specialty. The standard rotor provided optimum capacity for most corn and wheat conditions. Hard-to-thresh crops and tough-stemmed soybeans required the specialty rotor.

Underneath the rotor, the patented cross-flow cleaning fan delivered uniform airflow across the sieves. The chevron fan-blade was the key to this design. The full-width frontal fan air intake increased the fan's efficiency.

The grain unloading system had a 10 percent increase in grain auger speed and new controls to make unloading more productive.

In 1996, the 2166 and 2188 were fitted with Tier 1 emissions-rated engines. The 2144 was fitted with a Tier 1–rated engine in 1997. The 2100 Series engines were manufactured with the same technology used in similar models built for the Magnum tractor series. These CDC-built engines (Consolidated Diesel Corporation) offered plenty of reserve power and season-after-season durability.

In 1996, the 2100 Series received a new hydraulic system. The open-center (constant running) pump was replaced with a new closed-center system called the PFC (pressure flow compensated). The PFC system's hydraulic pump ran only when a hydraulic demand was placed on the system. The PFC system operated the steering, reel, unloader auger swing, header raise, and Field Tracker header tilting. This saved fuel and also allowed the combine to operate at a higher-rated system pressure. The extra pressure allowed the greater header lifting capacity needed to handle the larger headers.

Also in 1996, a third strand of chain and staggered slats were added to the 2188 feeder chain. This improvement increased chain-service life and helped reduce the vulnerability of slat-bending in severe conditions.

In 1997, the three main hydrostatic tubes from the pump to the motor were replaced by hoses with a higher pressure rating, allowing six tube joints to be eliminated.

The use of front-axle dual wheels became more common with the 2100 Series. As headers grew ever larger in size and farmers continued to add grain bin extensions to increase carrying capacity, a better way to combat compaction was needed. The use of row crop–spaced duals also allowed for less tire damage from field stubble. Today's corn hybrids are bred to have stalks more resistant to rot. This has caused a significant issue with serious tire damage and destruction; running spread duals instead of super-wide singles has saved thousands of dollars in tire damages.

Those 2100 Series machines factory-fitted with duals also had heavier final-drive housings and larger output shafts to handle the extra stress of duals.

The 2100 Series came standard with a new automatic feeder cutoff feature. This sensed the speed of the feeder driveshaft; if the speed dropped dramatically, the auto feeder cutoff stopped the header and feeder, protecting the combine from damage and downtime.

Several new safety features were added in the 2100 Series as well. A new hydraulic disc parking brake, for example, was engaged simply by pushing a rocker switch. In addition, an operator presence feature ensured that the feeder and separator systems were off when the operator left the cab. If the operator left the seat for more than seven seconds, the operator presence feature disengaged the two systems.

Case IH 2188

▲ Frontal view of the 2188 fitted with a 1063 corn header, a six-row head set to harvest 30-inch-wide rows. Owned by Andrew Tucker. *Lee Klancher*

Another safety feature was additional screen shielding on the lower sides of the combine. These perforated screen shields could be unlatched and opened to allow full access to the machine for adjustments and servicing.

The use of wide marker or extremity lighting on the header attachments also began with the 2100 Series. This locating of warning lights on the header extremities helped others to see the unit's width during nighttime travel.

Despite all these new features, the 2100 Series was short-lived, with only a three-year production run.

NEW CAB FOR 2100 SERIES

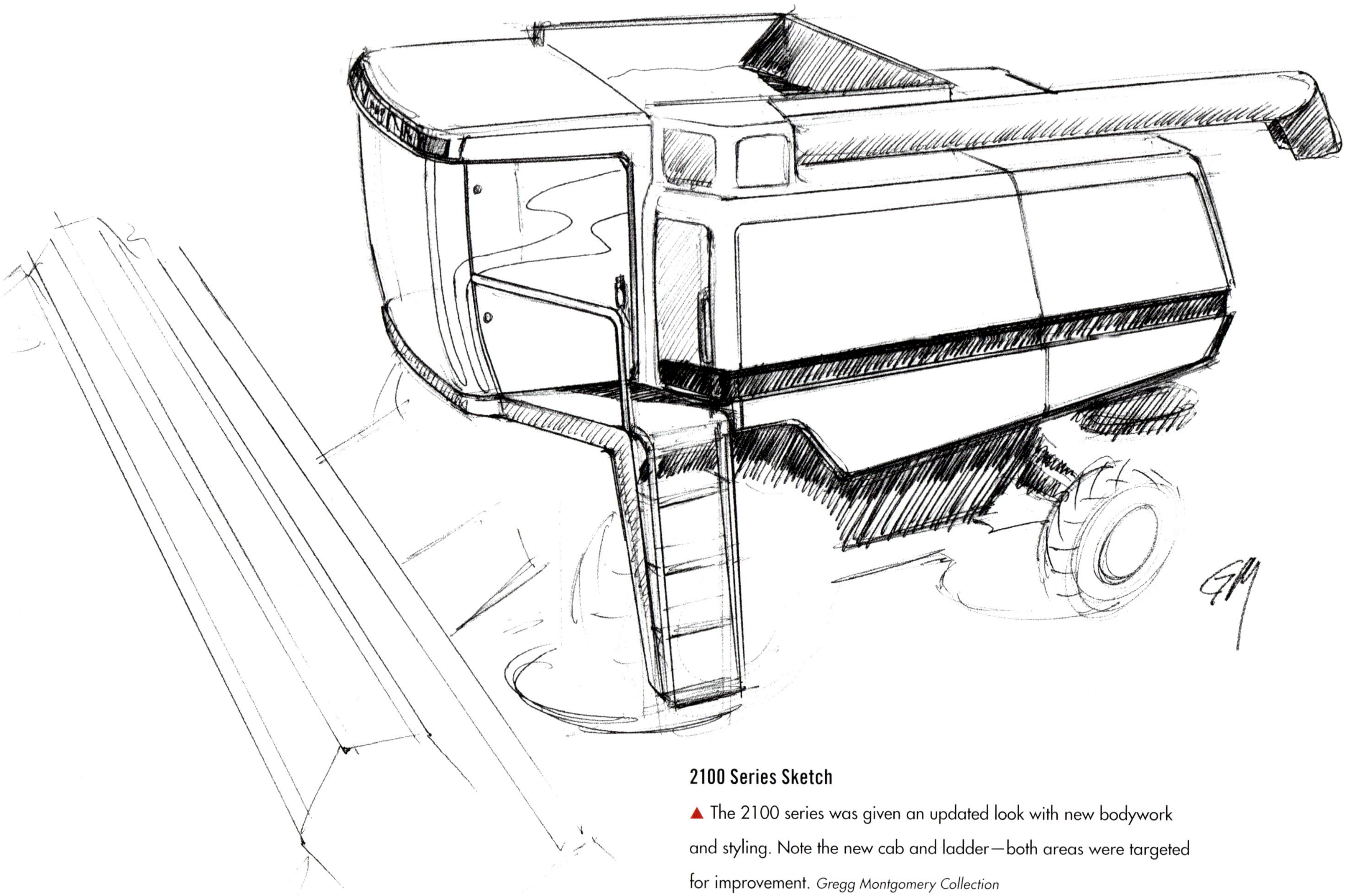

2100 Series Sketch

▲ The 2100 series was given an updated look with new bodywork and styling. Note the new cab and ladder—both areas were targeted for improvement. *Gregg Montgomery Collection*

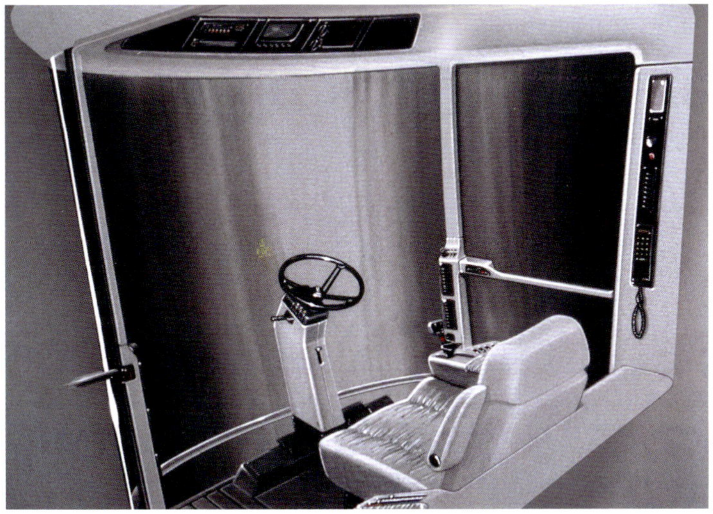

New Cab Design

▲ The new cab developed for the 2100 Series.

Gregg Montgomery Collection

Fitting in the Lab

◀ Rough mockups of the new cab were tested by staff at Montgomery Design International.

Gregg Montgomery Collection

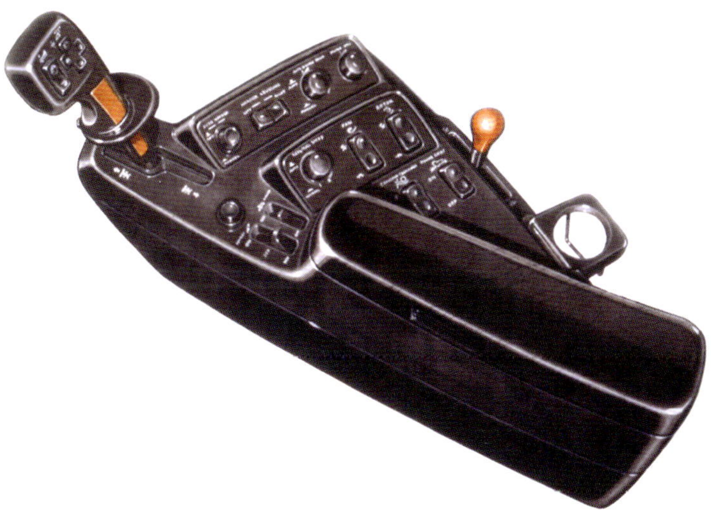

Control Console Sketch

▲ An all-new control console was developed for the 2100 Series' new cab. The clever hand controller would become the multifunction propulsion lever that allowed the operator to more easily control complex operations such as unloading on the fly.

Gregg Montgomery Collection

Production Cab

▲ The production version of the new cab was a major upgrade, featuring better visibility of the instrumentation and more intuitive controls. *Gregg Montgomery Collection*

2300 SERIES

By Ken Updike

In 1998, the 2100 Series was replaced by the 2300 Series. These combines featured a number of improvements..

The 2300 Series saw the Axial-Flow family expand to production at the Case IH factory in Brazil in 1997. There, the 2388 model was made for the South American market and for export to markets like South Africa. (The Brazilian Axial-Flow combines were not sold or imported into North America or Europe.) The only other time in Axial-Flow history that a combine was *not* made in East Moline was in 1979, when IH outfitted its Angers, France, factory to build Axial-Flows for the European market. These machines (besides having the required Euro safety lighting and markings) were near copies of the East Moline versions, even using many parts that were made in Illinois and shipped to France for final assembly. The major difference was in the combines' radiator air screens—the East Moline units featured a rotating screen that extended to the side of the combine, increasing its net width; the French-built Axial-Flows had the air screen mounted at the rear like the 1600 Series Axial-Flows did nearly 10 years later.

In 2000, the crop-harvesting marketing department at Case IH seized the day with its "Rotary Combine Challenge" marketing campaign aimed at Case IH's main combine competitor, John Deere, which had just announced its own rotary threshing combine (except that Deere was quick to stress that it was *not* a rotary combine).

A year or so after the Deere launched, Jim Irwin at Case IH decided it was time to respond. He gave marketing manager Kelly Kravig the green light to create a poster that would portray the strengths of their machine.

Kravig found a graphic design firm in southern Wisconsin. When they showed him their samples, he thought it looked familiar. It was—the artists worked for Disney.

2300 Series

▼ The 2300 Series combines (the 2344, 2366, and 2388) are shown in a group photo. The 2344 was produced for four years before it was retired from production. *Case IH*

Cutaway

▼ The internals of the 2300 Series combines. All Axial-Flow combines shared the same basic design in their threshing and grain-cleaning systems. *Case IH*

Rotary Poster

▶ This poster was drawn by a Disney animator and released when Deere and Co. finally introduced a rotary combine. The poster was so popular it was reprinted in 2014. *Case IH*

Case IH 2366 100,000th Edition

▼ In 1996, Case IH built the 100,000th Axial-Flow combine at East Moline, Illinois. It was a specially striped 2366 model. *Case IH*

Kravig spelled out that they wanted to see fat, happy birds behind the Deere, and skinny, starving versions behind their machine. The resulting piece of art was released as a limited edition poster.

"When we released the poster, it was extremely popular," Kravig said. "Normally, you do one run and never do another one. We did a second and a third printing, and just launched it again in 2014."

Combines operate in dusty, hot conditions, and the part that seems to take the most abuse is the alternator. With more electrical systems added, larger alternators were installed to meet the demand. Of course many owners added extra lighting for better nighttime visibility, which only increased the electrical load. Since heat and dirt can quickly combine to kill an alternator, keeping the alternator clean and cool has always been an issue. The original 1400 Series had an effective

dust screen. The 2300 Series had an improved dust screen with a vacuum system that used engine fan air to keep the alternator cool and running.

It is worth noting that the "40" size combine (1440-1640-2144) was retired when the 2344 model stopped production in 2003, the reasoning being that the used combine market offered more affordable machines with the same or greater capacity than a new one could be built to accommodate. The market demand for newer, smaller machines was declining.

In 2001, the Rochelle Rotor became available in the 2300 Series. Developed by Walker-Schork International, a Case IH dealer from Rochelle, Illinois, the rotor was a variant of IH's specialty rotor design. Also in 2001, the 2388's fuel tank capacity was increased to hold 180 gallons of fuel.

Also new with the 2300 Series was a factory-installed Hillco Technologies hillside leveling system that allowed the Axial-Flow to traverse steep hillsides. (In the past, combines were shipped to Hillco to be converted.)

Case IH 2388

▲ This 2388 makes short work of tall corn. *Lee Klancher*

Case IH 2388

▶ Harvesting rice is extremely tough on any combine. The simple design of the 2388's threshing system allows the operator to make adjustments from the cab as crop conditions change. *Case IH*

Case IH 2388

▲ This design drawing shows the flat upper-rear body panel design of the proposed AF 2388 model. *Gregg Montgomery Collection*

Case IH 2388

▲ Another design drawing shows the more stylish rounded rear panel design and bodywork. *Gregg Montgomery Collection*

In 2002, the 2344 and 2366 received a 130-gallon fuel tank. Also that year, one-piece concaves were offered on the 2300 Series.

In 2005, the 2377 replaced the 2366. That year also marked the move of combine production from the old East Moline Works to the new Case IH factory in Grand Island, Nebraska, which originally had been built in the 1970s to produce New Holland's self-propelled combines and forage harvesters. As of this writing, the facility continues to build Axial-Flow combines for North America, Europe, and Australia.

After the Case IH and New Holland merger in 2000, both companies kept their combine production separate, though the engineering departments were combined with the eventual goal of creating a combine that shared most parts but whose design was specific to Case IH or New Holland in the threshing/separating area. CNH still uses this "common platform" engineering approach today in combine design.

Post-merger it was decided that building both brands of combines in one factory would be the best for the company in the long term. The East Moline Works was demolished after combine production moved to Grand Island. The last 2366 built was serial number JJC0257696. The final 2388 built at East Moline was serial number JJC0276540 in 2005. Coincidentally, the 1,000th Brazilian Axial-Flow was built the same year.

The 2300 Series' eight-year production run ended in 2006, but not before the location of the serial number plate was changed from the left-hand main frame rail to the front side of the right-hand deck support.

Case IH 2377

▲ In 2005, the 2366 was replaced by the 2377. The same year, the AFX rotor used on the 8010 and 7010 was installed in the 2388 and 2377. *Case IH*

MAIN STREET TOUR PROGRAM

In the late 1990s, new Case IH executive Jon Carlson came to marketing manager Kelly Kravig with an unusual proposal: Carlson asked Kravig to cut four combines in half.

At the time, Carlson, son of "Tractor" Bob Carlson and a former John Deere executive, was working with the management team to revitalize the combine line. The critical piece of that freshening would be the new AFX machine, but it would not be introduced until 2003.

In the meantime, the team looked to strengthen its position by marketing the existing line and building customer relationships. Gerry Salzman recalls that was accomplished by having product marketing and sales directors (including senior management) get on a plane to visit customers and dealers.

The other portion of the strategy was the Main Street Marketing Tours, which put demonstration exhibits on semi trucks traveling around the country and stopping at Case IH dealerships to put on presentations. The first of these was the Magnum Showdown Tour in 1996, which proved a popular success.

In 1997, Case IH brought the Magnum tour back and added three more tours: the Steiger Specialty Tour, the AFS Precision Farming Tour, and the Axial-Flow Advantage Tour.

According to Salzman, Carlson wanted to go directly to the customers with a professional sales story backed by facts and enthusiasm. For the Axial-Flow tour, Carlson wanted red and green combines cut in half so customers could *see* differences.

Kravig couldn't literally cut the combines in half, but he could build cutaway machines that exposed their vital parts. He and a handpicked team did just that and added electric motors to run the systems and lighting needed to show it all off. This was accompanied by a movie that demonstrated how all the features worked.

The tour was assembled on semi-trucks that rolled out of Phoenix, Arizona, to more than 120 dealerships around the country. Dealerships drew anywhere from fifty to four hundred people to each event. Participants were given medals and thousands of farmers came away understanding the Case IH machine.

Road Show

▲ The Main Street Tours traveled around the United States in 1997, showcasing new Case IH equipment. *Case IH*

AFX SERIES DEVELOPMENT

By Lee Klancher

"I cannot stress too much the overwhelming importance of technical change as a primary force that will likely be reshaping farm supply conditions—as it has been doing for a long time." —Alan Greenspan, in a March 16, 1999, speech

In the late 1990s, Jay Schroeder was working in Burr Ridge when a supplier mentioned that New Holland had asked for a quote on a similar, critical combine component rated to handle 425 horsepower.

"I have no idea what New Holland is doing, but that's a really big combine," the supplier told Schroeder. At that time, the biggest combine on the market had a 280-horsepower engine. Case IH was working on the CBX, a larger 600-horsepower machine.

Schroeder later recalled hearing that while the IH engineering group was working on the original Axial-Flow in July 1972, it had discovered a prototype New Holland rotary combine at a truck stop in Illinois.

He was struck by the similarity of the situation. "Here Case IH is working on CBX and we find that New Holland is about a year or so ahead of where we are," Schroeder said. "I thought, '*Boy, this is not a good cycle*.'"

The cycle was about to be broken.

By the spring of 1999, the good times the farm economy had experienced in the middle of the decade dried up and blew away. The American economy was relatively strong, but the demand for exports dropped drastically due to depressed economies in Japan and Russia, as well as decreased demand in other Asian countries and South America.

Corn prices dipped to $2 a bushel in the Midwest, the lowest level in a number of years. Soybeans, wheat, hog, and dairy prices all took a tumble as well.

The effect on combine sales was predictable—sales reduced by roughly 50 percent industry-wide. As the bottom dropped out of the market, a new partnership emerged that

8010 AFX Design Sketch

▼ The AFX 8010 represented a very radical change in design for the Axial-Flow. Generally, the only part carried over from the traditional Axial-Flow was the single-rotor design. The concaves, sieves, feeder, grain tank, drives, cab, and engine were all larger and much different.

Gregg Montgomery Collection

would rock the world for several combine development teams.

In May 1999, Case Corporation and New Holland N.V. announced the intention to merge. Fiat S.p.A. in Italy owned a majority of New Holland and would purchase Case Corporation and merge it with New Holland, creating a new company.

Jean Pierre-Rosso, previously CEO of Case Corporation, would be the CEO of the new company, which would have a combined revenue of more than $12 billion. Deere & Co. revenue in 1998 was $13 billion—the merger created a company that was within striking distance.

While executives of all the agricultural companies fretted over competition, the fans of red combines worried about color.

In late May 1999, radio announcer Max Armstrong talked to the *New York Times* about the issue. "One guy called in yesterday; he was all lathered up about it," Armstrong said. "It's just too weird to imagine a Case IH combine any color but red."

Armstrong grew up on a farm and spoke of schoolyard fights between those who loved red and those who favored green. He wryly added, "Of course, if they weren't red or green, we just felt bad for those poor kids."

Bill Masterton of Case Corporation assured the *Times* that the Case IH and New Holland brands would live on. That turned out to be true, and eradicating one of the brands doesn't appear to have been seriously considered as part of the merger.

Perhaps the memories of the struggle to create a new brand identity when Case IH was created in the mid-1980s were strong enough to dissuade anyone from trying to repeat the feat.

But that isn't to say that things were crystal clear. The Case IH and New Holland brands would live on, but product lines and facilities needed to be merged in a rational way. Plants would close and jobs would be cut; determining which ones made sense to keep was a difficult job.

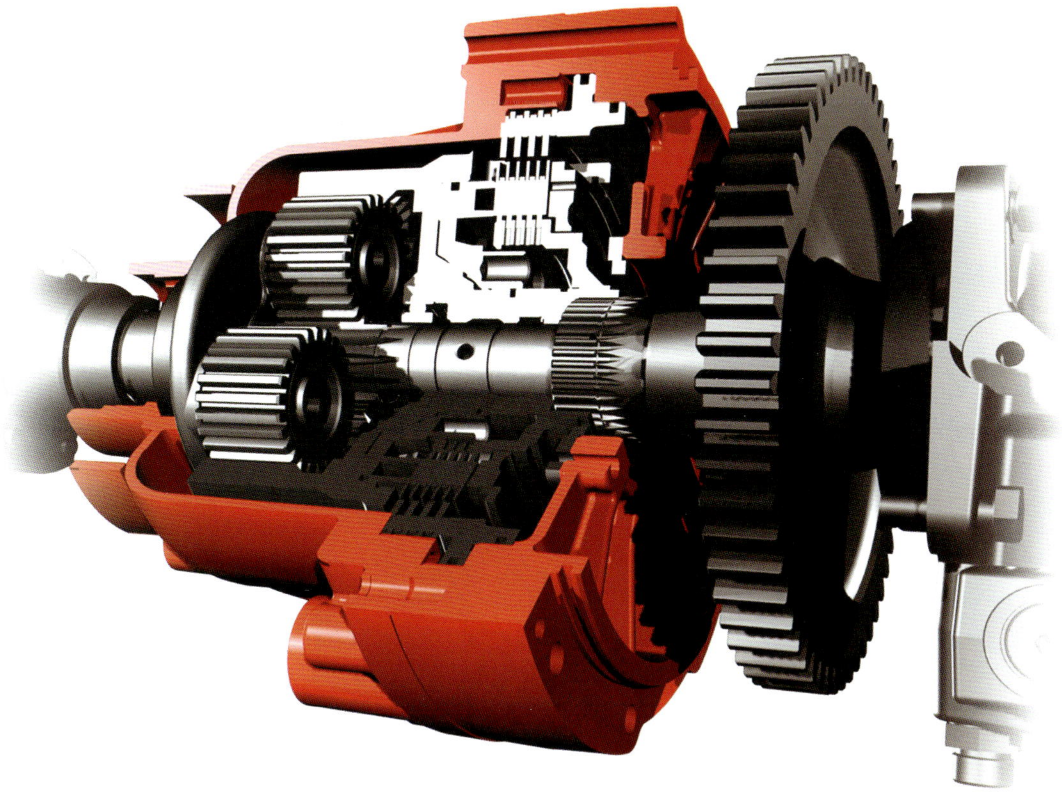

AFX Series CVT Drive

▲ The new AFX series combines featured an innovative hydro-mechanical CVT drive for both the feeder and the rotor. *Case IH*

Market by market, the management team at Case New Holland had to determine which pieces of equipment lived on, which lines were to be consolidated, and which were to be discontinued.

For the men and women whose livelihood revolved around combines, this was a particularly tough time. They had come through the Case IH merger relatively unscathed. The IH combine business had been strong, and J. I. Case did not have a combine line so the decisions were fairly straight forward. That was not true this time around. After nearly a year of intense research and analysis by the global product marketing teams from Case IH and New Holland, proposals were presented to management. Given the strength of both brands in combine harvesting, it was determined that the best option was for the Case IH combines to stay with the single rotor Axial-Flow system, and the New Holland machines would keep their twin-rotor design. This wasn't the most economical decision, but it was the right one for the brands.

2004 AFX Cutaway View

▲ The 8010 offers a high-capacity 330-bushel grain tank and the ability to unload three bushels per second. The 2062 Draper Header is 36 feet wide and can harvest 24 acres per hour. The AFX combines were updated extensively for 2006. *Case IH*

THE COMPETITION

While Case IH was developing the CBX in the late 1990s, New Holland was working on the CX/CR combine family. The CX would have a conventional threshing system, and the CR would receive a twin-rotor system. The New Holland design was bigger than what existed on the market, but smaller than the mega-sized CBX.

"At that time New Holland had a very outdated combine product line," New Holland engineer John Hansen said. "The company was at a point where they needed to decide whether they were ready for future North American business, going to reinvest in the product line and start to develop a complete new combine, or get out of the business."

John Hansen grew up in the Quad Cities area of Illinois and Iowa, surrounded by the sons and daughters of people who worked for Case IH or John Deere. He went to college to study mechanical engineering and described going to work in the tractor industry as a "natural thing."

Hansen accepted an engineering position with New Holland in New Holland, Pennsylvania, working on tractors and transmission design, and in the mid-1990s moved to combine driveline design. Like Lucas and the Case IH engineers, he was pleased to be developing a brand-new high-horsepower machine.

"Luckily for me, they decided they were going to make a big investment," Hansen said. "I was at the very start of a new combine platform development where we basically started with the clean sheet of paper."

The company considered adapting the bi-rotor design, an innovative rotary threshing system developed by two cousins from central Kansas, Mark Underwood and Ralph Lagergren. Underwood had installed his design in a 1480 combine known as "Whitey" due to its white paint, and Lagergren took on the financing and marketing of the machine. The two men were doing all they could to sell their concept to a major manufacturer and shopped it to everyone in the market, including Case IH. New Holland and Caterpillar were considering a joint venture to adapt the pioneering design to their technologies.

The technology was promising but development costs were high, and New Holland determined it was better to proceed and build its own clean sheet machine. Deere & Co. eventually purchased the rights to the bi-rotor and, as of 2015, the technology has not been seen on a production machine.

The New Holland design of its new combine platform proceeded. An early concept was the modular design that allowed multiple platforms to be pieced together.

That modular design concept would prove critical to both future red and blue combines.

TOUGH DECISIONS

The CNH merger required the two combine development groups to join forces and devise a way for the two brands to survive independently yet maximize efficiency. The legacy New Holland combines were focused on small grain harvest, while the Case IH legacy combines were focused primarily on corn and soybean harvest. The existing machines were complementary.

The large combines in development were different. The Case IH CBX was a giant, designed to tackle Class 8 and way beyond. The machine offered a high level of innnovation, and it also required a very large investment in dollars and time to complete.

The New Holland machine was larger than anything on the market at that time, but designed to slot into Class 6 to 8. The machine was also further developed and could be brought to market in a shorter amount of time with less dollars. How to move forward on this front was a very difficult decision.

"The AFX program management team under the leadership of Rich Holloway, platform manager, put the New Holland and Case IH engineers together, and we compared notes on what's the best based on our experience, what were the best things we collectively had in hand, and what are those things that we don't have a good solution for and we really need to invent a better solution," Hansen said. "We basically shared the technologies we had been working on. I was particularly focused on drives at that time. It turns out that both New Holland and Case IH were in parallel pursuing CVT (hydro-mechanical) drives but going about them in just a little different way."

The decisions regarding how the combine lines would be merged happened slowly. A variety of options were discussed, and plenty of long and not entirely peaceful meetings were held to make a choice. The process of deciding what to do with the new flagship combine business took nearly six months.

PROJECT GREEN SWEEP

While these decisions were being worked out within CNH, the competition saw vulnerability and struck.

In June 1999, a John Deere advertisement in *Farm Industry News* displayed the rear end of a concept combine and the tag line, "A new era of harvesting is taking shape." Speculation was rampant. In August, Deere and Co. announced its new rotary combine. The market was still soft, and Deere's approach was to introduce a new product line and cut prices mercilessly.

In addition to its public campaigns, Deere gave its sales force a pricing/promotion scheme called "Project Green Sweep" that offered lower-priced leases, low-interest financing, rebates, and high trade-in allowances for the Case IH customer who purchased Deere products.

The program was described in a June 4, 2000, article in the *Milwaukee Business Journal* titled, "Deere Price-Cutting Plows Some Case Dealers Under."

The incentives worked, and Deere was able to wrangle a number of Case IH customers to take the cash and attractive terms and put new green combines in their sheds.

Deere's net income rose 36 percent in the first fiscal quarter of 2000, and CNH reported a loss while workers idled in the plants in Racine and Fargo. Despite the results, Case IH refused to price match.

Jim Sommer, the owner of an Appleton, Wisconsin, implement dealer, saw a 36 percent reduction in his Case IH sales in the first quarter of 2000. When speaking to the *Milwaukee Business Journal*, he didn't pull any punches.

"The time it is taking the companies to get together is detracting from sales and marketing," Sommer said of CNH and Case IH. "Deere is keeping their factories humming. Case is closing factories."

BROAD DIRECTION

The pressure was on to find a way forward, and in November 1999 it became clear that the new Case IH Class 8 combine would use New Holland's CR design as the base platform.

One key factor in the New Holland design was that the machine had a modular design.

Don Watt, engineering program manager for the Case IH AFX program, explained. "They split the combine, top to bottom, so there's a lower chassis and an upper chassis," he said. "You had a narrow and a wide upper chassis and a narrow and a wide lower chassis so you could have a mix and match." This was done to allow large and small combines to share componentry.

That modular design was critical when New Holland and Case IH were jointly aligned under the CNH banner.

"We were both innovating in that area and really had no idea what the other company was doing," Hansen said. "When we really compared our experiences, it gave us an opportunity to leverage our individual experiences and determine what worked best for both brands."

The modular design offered them a number of key advantages. For one, the machines could be upgraded one module at a time. A new cab could be created, for example, without having to revise the entire platform.

And more importantly at the time, the two development teams could create their own versions of different systems to suit the brand DNA requirements. This allowed the company to build two different machines on the same production line and effectively create products that had different advantages to suit the market.

"One benefit I saw was the cooperation and the ability to use the knowledge of both brands to develop products," Hansen said. "I saw a big collaboration that allowed us to learn more about the market."

The market they were serving was also more global than ever, which magnified the complex challenge of building a combine that worked in vastly different conditions.

"Probably the one thing on the combine that makes it relatively adaptive for a lot of conditions is that it's a modular build," Jim Lucas explained. "For example, there's an engine module, a lower chassis, a threshing module, grain tank module, those types of things that makes it . . . easier to make changes to certain

2004 AFX Cleaning System

▼ An all-new cleaning system debuted on the AFX combines. This portion of the combine saw a lot of development work in the 1990s and 2000s. *Case IH*

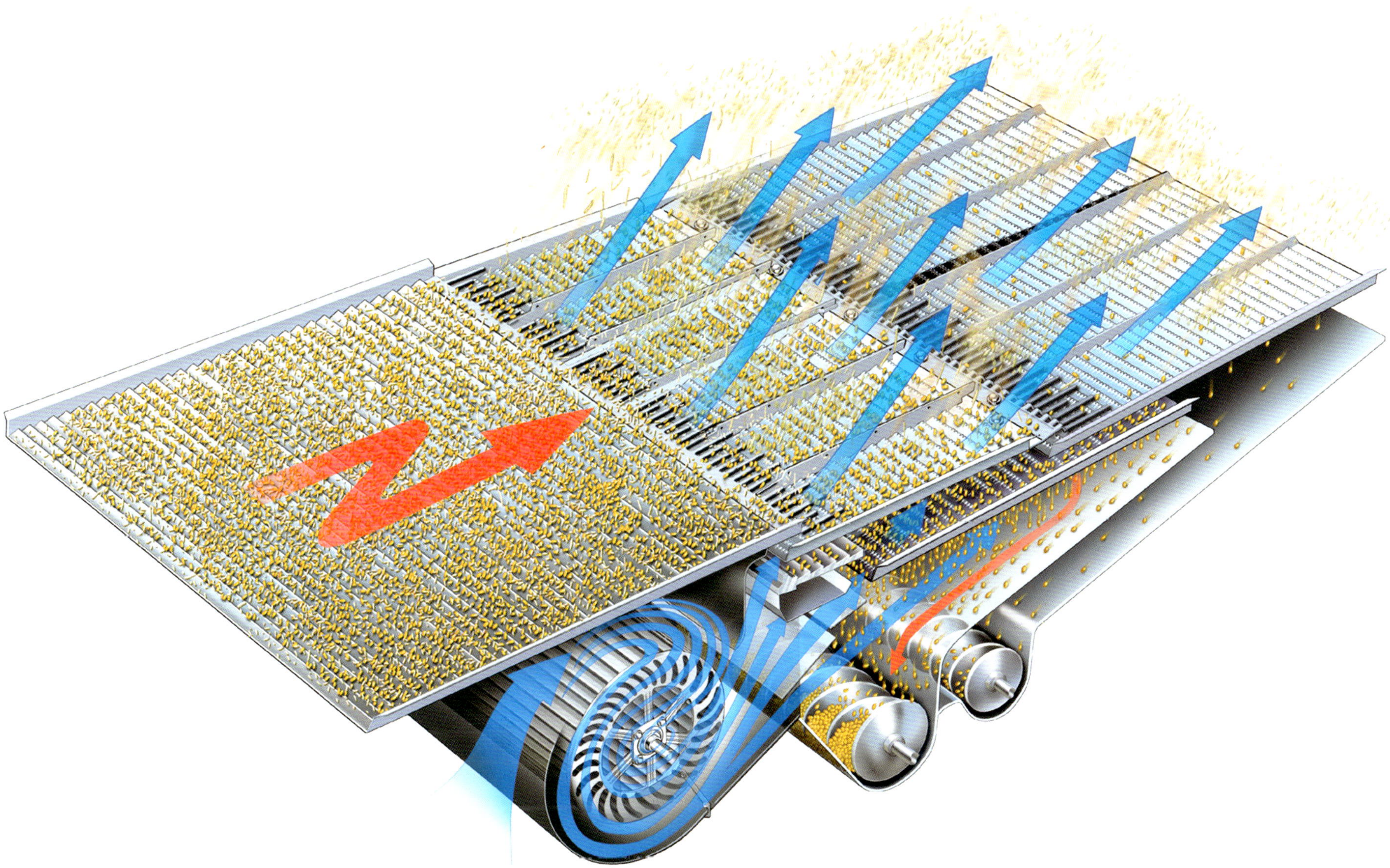

parts of it without having to completely redo the whole combine and/or adapt different modules for the different markets we have, whether it's here, Europe, South America, or Eastern Europe."

The split chassis also allowed the Case IH and New Holland machines to retain their unique features and share componentry. The question was: Would the single rotor work in the CR chassis?

"The first thing we had to do was sit down in New Holland, Pennsylvania, with a couple guys who were pretty good with CAD systems to determine how could we fit a single rotor in this CR frame that New Holland had," Watt said. "Bob Matousek and I sat in a darkened room with several of the key New Holland combine engineers using Computer Vision, which was an archaic software tool, as the New Holland organization had not yet migrated the design process to 3-D solid modelling, but that's what New Holland used, and we tried to figure out how big a rotor (both diameter and length) we could actually put inside this frame."

The space was shorter than they hoped, but they were able to increase the threshing area and increase performance. Nearly a year after the merger was announced, the blended engineering teams saw signs that both could introduce new flagship combines in their respective brands' DNA and continue providing their customer bases the high capacity, reliable combines they were expecting.

Similar processes took place with engines and cabs and controls and all the myriad of other components that would make up the new flagship Case IH combine. Progress was being made, and the design and development was picking up speed. The dealer and customer community, in the meantime, was nervous.

Gerry Salzman had been part of the task force assembled to develop the product strategy for the Case IH combines. In February 2000, Salzman, Sam Acker of product marketing, and engineer Don Watt traveled to the New Holland technology center in Pennsylvania to present their ideas to a half-dozen key dealers (including Charlie Hoober).

In that meeting, they outlined the brand strategy for the Case IH and New Holland combines. They reassured dealers that the Case IH combines would retain the single Axial-Flow rotor, and New Holland would continue to have the twin-rotor design. Similar meetings were held around the world to confirm dealer support for the plan.

The reaction was relief and support with a strong request to move quickly but most of all . . . "Do it right!"

CREATING THE AFX

Once the platform was decided upon, the team had a mountain of work to do to assemble all the parts and the pieces that would allow the Case IH DNA to remain strong in the new AFX.

"The AFX development process would be a classic example of Platform Engineering," said Don Watt, engineering program manager.

From a technology perspective, elements and systems from both the Case IH CBX program and the New Holland CX/CR platform were used in the AFX development. Several completely new designs were developed as follows:

- The Case IH Axial-Flow single rotor threshing and separating concepts were used.
- The New Holland cleaning system was adapted.
- The Axial-Flow traditional cross-flow cleaning fan system was incorporated.
- CX/CR modular split upper and lower frame main concept was utilized.
- The CBX hydro-mechanical transmissions for rotor and feeder/header drives were adapted to the AFX along with the concept to replace most belts and chains with small gearboxes.
- A new AFX high-capacity vertical reside management system was developed.

- The CR/CX grain tank and unloading concepts were used.
- The AFX clean grain handling system was developed to provide significantly higher capacity in corn harvesting.
- The AFX hydraulic system was an adaptation of the CBX system.
- Engine systems from the CX/CR were incorporated with traditional Axial-Flow air management and cooling systems.
- The cab structures were common between the AFX and the NH combines with each version equipped and trimmed in the brand DNA to provide the expected brand customer interfaces.
- The CX/CR electronic system was adapted to AFX architecture using AFX-specific wiring schemes.
- The AFX combine was the first in the industry to use the display to manage combine functions.

Developing all these new systems would take time—too much time, as far as company leadership was concerned. Deere's Green Sweep program and the slow sales of combines put pressure on Case IH to release the new model as soon as possible.

The development team wanted four years and additional funding to complete the AFX. The executive decision was made that the market demanded the machine be out sooner, so the time to introduce the new machine was mandated to be three years, launching in 2003.

As history has shown, rushed development tends to end up with predictable (and unfortunate) results.

"It was a push to launch AFX, and the guys were working intensely," Jay Schroeder said. Schroeder's work on the rotor drive would pay off, as that technology was adapted to the new AFX combine.

"The hydro-mechanical drives were initially developed in the CBX program," Jim Lucas said. "We designed the hydro-mechanical drives for both the rotor drive and the header drive in the CBX program and they were carried over onto our flagship combines, the AFX 7010 and 8010."

The AFX combines were the result of two engineering teams' work, and their modular concept would allow red combines to continue to be a unique machine. In this instance, at least some of the work done creating the remarkable CBX design would live on.

The CNH challenge was not only to develop the AFX family of combines close on the heels of the New Holland CR launch, but to bootstrap the former New Holland Grand Island, Nebraska, low-volume combine plant into a world class, much higher-volume combine production operation. In addition to managing the influx of an extremely large number of new components and assemblies to support the AFX and CR combines, the Grand Island team faced the equally huge challenges of basically gutting the plant and installing all new equipment, and developing new manufacturing, assembly, test, packaging, and shipping processes, including a new major paint system. It was an enormous undertaking on a much-accelerated timeline.

MOVING THE MIDRANGE COMBINE

By Lee Klancher

By the late 1990s, the East Moline plant had 2.4 million square feet, and a good portion of it wasn't being used. The plant was as efficient and cost-effective as ever, but it was old and under-utilized, with ceilings that were too low to build larger combines and line equipment that was aging.

Steve Horst was a manager at the plant at the time. He also recalls how hard they worked to keep their plant alive despite the challenges.

"The plant put up a valiant effort," Horst said. "There were a lot of plans that were drawn up, proposals that were made. There were a lot of meetings that were held."

The East Moline plant was too old, small, and unwieldy to survive. In 2000, it was announced that the plant would close in 2004.

Steve's brother Jamie was another key member of the East Moline plant, helping with quality control and logistics during production moves. He recalls the period from 2000 to 2004 with pride.

"We performed like there was no tomorrow. We didn't miss a schedule," Jamie said.

The plant was asked time and again to meet the growing demand for Axial-Flow combines and other equipment. When asked, they could add fifty or even a hundred combines to the monthly run.

"I can't remember anything not ever being met, not being done that they wanted," Jamie said. "We knew the machines, we knew the product, we knew the staff, and it was good."

Once the decision was made to close the plant, the East Moline staff began moving production lines to their new locations. Head production was moved to a plant in Saskatoon, Saskatchewan. Cotton pickers would be built at the Benson, Minnesota, plant.

The fate of the midrange combines was heavily debated. The 2300 Series machines were a highly evolved version of the original Axial-Flow introduced in 1977. Part of the management team believed that the series had run its course and all the focus should be put on AFX combines. Others saw the need to offer smaller machines with solid features and an attractive price. The North American sales and marketing group looked at it differently as they felt a real need to offer the proven 2300 series with solid features and an attractive price. Heated discussions took place regarding the line's fate.

In the end, the midrange line was saved and production was moved from East Moline to the Grand Island, Nebraska, plant, which was already building the AFX Series machines. Grand Island provided affordable, high-quality combines for thousands of farmers.

Steve Horst was asked to help move the line. They told him to select a team of twelve for the job. The team's task would be to transfer the line equipment and production knowledge to the plant in Grand Island.

On August 20, 2004, the East Moline plant produced its last combine.

Steve and Jamie Horst, both near retirement age, were offered positions in the new plant, helping to get the line running. They and a few others took transfer positions.

On Labor Day weekend 2004, seven of them drove to Grand Island.

Steve Lee, retired Grand Island manager, recalls that the East Moline people were critical to making the transfer a success. "We had a very good complement of the East Moline expertise that came to Grand Island," Lee said. "Their experience was invaluable to us."

The changes required to move production of the 2300 Series combines were tremendous. The biggest was that they would be building Case IH AFX and 2300 Series combines as well as the New Holland combines on the same line.

The manufacturing team was also dealing with a dramatic reduction in space, from 2.4 million square feet in East Moline to less than 1 million in Grand Island.

They also had to ramp up production at Grand Island. The plant had been building four to five New Holland combines per day. The East Moline plant could crank out twenty combines a day. Grand Island's line output would have to be dramatically increased, in less space and with two different lines of machines.

One change that proved beneficial was the way paint was applied. At East Moline, combines had been painted after assembly. The Grand Island plant had a new system that did the painting prior to assembly, which was great. The issue was the capacity—that system could not handle 20–25 combines per day.

Last 2388 Built at East Moline

▲ With the East Moline Plant closing, production of the 2300 Series moved to the combine plant in Grand Island, Nebraska. *Case IH*

Steve Tyler was the senior director of manufacturing for Case IH from 2004 to 2006. His job was to oversee the transition from East Moline to Grand Island, and he recalls the paint system as one of the toughest challenges of the process. In the end, a large investment was made to upgrade it. That change proved positive for the midrange combines, as the paint applied after assembly on the East Moline line had hidden flaws. Quality improved and the machines looked sharper with unpainted hoses and fittings.

Other changes included the creation of a new logistics system that ensured the proper parts were delivered to each workstation. This meant creating new carts that could accommodate all of the combine lines and outsourcing hundreds of parts. "The Grand Island team . . . grabbed the suggestions and ran with them," Tyler said. "Made it all happen."

For Steve and Jamie, the jobs in Grand Island were transfer positions with a timeline. Both would leave once they felt they had the plant running smoothly. And both hoped their efforts brought a little East Moline attitude into the new location.

AFX 8010

The AFX 8010 combine launched in 2003. The basic design was cutting edge, and the machine set new standards in capacity and productivity.

The machine, however, was not fully developed and launched with a variety of issues that still needed to be resolved.

Terry Wolf from Homer, Illinois, is a longtime Case IH customer who has seen his business improve due to the efficiency of rotary harvesters. An active member and former chairman of both the Illinois Council on Food and Agricultural Research and the U.S. Grains Council, Wolf was honored as an Illinois Master Farmer by *Prairie Farmer* magazine in 1999. He's a leader among his peers and also a progressive farmer eager to try out the latest technology.

Wolf was one of the farmers who tested the early prototypes of the AFX combines in the early 2000s. When the 8010 hit the market in 2003, he bought one of the first machines.

The 8010 didn't work terribly well for Wolf.

"It was a pretty trying experience," Wolf said, "but we learned a ton about the machines. I learned a ton about the company."

The 8010 was constantly experiencing issues, and since the machine was brand new, parts and service support weren't always available.

"The whole engineering and marketing teams were closely following all the combines to see what problems developed," Wolf said. "I am sure they never expected the level of problems they encountered, but they worked hard to solve them. Since it was brand new, there were not a lot of parts available, and when problems occurred, they were trying to test new changes in parts, which took time as well.

"It was a very frustrating experience for all, but at the end of the season, they had solved a tremendous amount of problems, found solutions, and were ready to work that winter

Case IH AFX 8010

▶ An AFX 8010 fitted with a draper header harvests wheat. The AFX 8010 was introduced in 2003 and has a 375-hp, 10.3-liter engine. The 2003–06 models had tops, grain tanks, and other components painted black. The 2007–08 models would receive a color scheme with more red. *Case IH*

Case IH AFX 8010

◀ A black-top 8010 harvests corn with a 2412 header. The AFX features a self-leveling sieve. The red side panels are made of a composite material that won't dent or rust. *Case IH*

Case IH Benchmarking

▲ In July 2004, a Case IH team took to the fields in Arizona to benchmark the performance of its combine line. The results showed the machines had significant advantages over the competition. *Case IH*

Upgrades

▲ Earlier AFX combines had a number of issues that proved to be a setback for Case IH combines. The engineering team dug in, found fixes, and released dozens of improvements. By 2007, the bugs were fixed, allowing the combines' exceptional performance to shine. *Case IH*

Computer Technology

▲ Increased computer technology improved efficiency and offered access to Advanced Farming Systems (AFS) technology, which tracked planting and yield by location. *Case IH*

to develop the first full-scale run of the combines. If they were to do this again, I suspect they would not sell customers the combines but rather have them run them as a beta test to prevent some of the frustrations at the customer/dealer/company level that occurred that year."

The improvement started immediately, and for 2004, the AFX released with a long list of improvements. Enough, as it turns out, to keep legacy customers like Wolf satisfied.

"The 2004 machine ran well, and they've improved them ever since," Wolf said. "They're just highly dependable now. You just seldom have any problems. It's a real testament to how they followed through on it."

Getting the AFX combine right required several years of work, and Hansen recalled hiring 40 to 50 new people to get the work done. "We had just a huge endeavor for about the next four years after the initial launch to solve every problem we could possibly solve on the AFX combine," Hansen said.

2007 "RED TOP" 8010 AND NEW 7010

By 2007, Case IH dealt with the issues, both on machines out in the field that were retrofitted and on the new machines. The 8010 combines from that year forward are easy to distinguish in the field due to the red grain tanks. The red tops were added to align the machines with the rest of the Case IH product line with more red and to be sure that customers could visually identify the much-improved new combines.

The 7010 combine was launched the same year and was a Class 7 machine that used the same platform as the 8010. Development of that machine required several additional years, and it came onto the market as a finished, solid, reliable combine that met customer expectations.

Project Green Sweep turned out to be less of a threat than it first appeared. During the first few years, the new John Deere rotary machine proved to have teething issues of its own, and Case IH was able to wrest back many of the customers who initially took the bait of low interest and high trade-in values.

The market rebounded, as agricultural markets historically do, and the existing line of Axial-Flow combines were selling well by the time the AFX was launched in 2003.

The process of creating the AFX was not an easy one, and Jim Lucas believes that the drive and persistence of his co-workers helped the company overcome the challenges they faced.

"I totally enjoyed working on combines and that's my love and my passion. I really enjoyed working with the people," Lucas said. "Case IH people…were always, I'd say, pretty good engineers. We could always solve any problem anybody gave us, so if somebody said, 'Okay, could you put a single rotor into this combine?' we did."

2007 Case IH 8010

▶ Case IH celebrated 30 years of Axial-Flow combines in 2007 and badged the 2007 model-year combines with the special decal seen on this grain tank. The 2007–08 8010 combines also received more red in the color scheme, with much of the grain tank and the apparatus at the top of the machines painted red. The grain tank is also larger, with capacity increased to 350 bushels. *Case IH*

Case IH 7010

▲ In 2007, the model 7010 combine with its 9.0-liter engine and 315-bushel grain tank was introduced. The combine was based on the 8010 and downsized to be a Class 7 machine. The Case IH draper head on this 7010 owned by Marty Nigon is ideal for harvesting soybeans. *Lee Klancher*

Case IH 7010

▶ Features included a 315-bushel grain tank with 24-inch foldable extensions and a 21-foot standard unloading auger. A 24-foot auger was optional, but both unloaded at a rate of 3.2 bushels per second. The AFS oval decal on the side of this 7010 indicates that it offers AFS Accugide auto guidance from the factory. The 7010 was built in 2007–08.

Lee Klancher

Case IH 7010

▶ This 7010 Axial-Flow harvests corn with a 12-row header and is fitted with dual-drive wheels. Like the bigger 8010, the 7010 had a self-leveling cleaning system and the CVT rotor and feeder drives. *Case IH*

THE QUADTRAC COMBINE

The rich soil of the Red River Valley contains a lot of sand and clay. When the big plains country gets soaked in rain, the fields soak up the water like a sponge.

For farmers in the region, wet weather can be troublesome. The crops dry more quickly than the soil, meaning corn or beans can be ready to harvest while the soil is still a mass of gumbo.

A fully loaded combine with a large head puts tremendous ground pressure on the tires. Wet conditions can make harvesting impossible.

After one particularly wet, difficult fall in the Red River Valley, Peter Christianson and David Meyer of Titan Machinery brought an issue to Gerry Salzman, Case IH Senior Director for Global Marketing. The Titan dealerships up north had customers who weren't able to get their crops out, even with the biggest high-floatation tires available.

The idea that they proposed was fitting a Quadtrac system to the AFX combine. "With the Quadtrac, we had the perfect system to put on," Meyer said. "Deere didn't have it, and a lot of the others weren't doing it. We thought this could be done fairly reasonably."

Case IH was open to the idea, and a small team was put together to assess the situation. A call was set each Friday morning with engineers from Racine, Burr Ridge, and New Holland, as well as several from the Titan product innovation group. Engineer Jim Lucas led the Case IH team.

Several issues came up quickly. First, the tracks would ideally bolt to the existing undercarriage, so they could be fitted on any wheeled AFX combine. They also needed to address ground speed issues; without some kind of gear reduction system, the tracked combine would have a top speed of about four miles per hour—too slow for transport.

"The team came up with the idea of changing the gears in the final drives," Meyer explained. The gearing change would give the combine an adequate road speed. Titan took the lead on fabricating the attachment of the tracks to the combine frame with a local supplier. The Case IH engineering team designed and fabricated the planetary final drives.

The process began in February 2007. By late June, the prototype was ready. In August, they started testing a unit out in the field. The Case IH engineering group optimized the track attachment structure to make it easier to manufacture and changed to a bull gear drive for the track system. By the following season, the the Quadtrac kit was available as a factory option. Demand was strong. Farmers around the country who had to deal with wet weather fitted the kits.

"This was an excellent example of working with a partner and existing technology to create a product that was needed," said Lucas. "I do not recall any other project happening that quickly."

"The floatation with tracks was outstanding," Meyer said. "They cannot only get out in the field and harvest, but they can work the ground without tearing it up and creating ruts."

Heavy Pressure

◀ As combines, headers, and grain tank capacities have increased, the amount of ground pressure on the drive wheels has also gone up. Machines like this Quadtrac-equipped 9230 and the Steiger 370 have lower ground pressure that allows them to harvest in tough conditions. *Case IH*

2500 SERIES

By Ken Updike

Case IH 2577

▼ In 2007, Case IH replaced the 2377 and 2388 with the 2577 and 2588. The 2577 Axial-Flow combine has a 230-bushel grain-tank capacity and an 8.3-liter CDC engine rated at 265 hp. The 2577 was built in 2007–08. *Case IH*

The venerable midrange Case IH combines were introduced along with the revamped 8010 and new 7010. The 2500 line was an upgraded version of the 2300 Series, which was an evolution of the original Axial-Flow platform.

The 2500 series Axial-Flows were introduced in 2007 and included the 2566–2577 and 2588 models. The new Tier 3 emissions rules had taken effect in 2006, so the series' 8.3-liter Consolidated Diesel Company (CDC) engines were fitted with advanced electronic monitoring with automatic shutdown. Both the 2577 and 2588 were approved for use with B20 biodiesel in 2008.

With the 2500 Series, unloading auger rate was increased to 2.4 bushels per second. The unloader tube length was increased to 18 or 21 feet to accommodate the use of larger headers.

The 2500 Series would be the last Axial-Flows built by Case IH before the combine model numbers were changed to recognize the class size of the machine they represented in a move meant to realign the combine models with a worldwide system. The new combines would be designated in size from Class 1 (smallest) to Class 10 (biggest). In today's models, the first digit in the model number indicates the combine size class. A 6130 combine, for example, is a Class 6 machine in the 130 series of combines.

With new model-numbering system in place and a new factory in which to build them, Case IH introduced a new combine that would set the bar for future models to follow.

Case IH 2588

▲ This 2007 model-year 2588 has a 2208 eight-row 30-inch-wide corn header attachment. The 30 years of Rotary Technology decal was affixed on all 2007 model-year Axial-Flow combines. *Case IH*

30TH ANNIVERSARY CELEBRATION

By Gerry Salzman

The Axial-Flow combine has had several of its milestone anniversaries celebrated with events, but the thirtieth was one of the events that really stands out.

Celebrated in 2007, the event rallied our company to the future around a critical part in our past. Held at the Grand Island Plant, the event brought together longtime customers, industry media, key commodity organization members, and senior management from Case IH, as well as CNH.

The event cemented the red combines in the hearts of the Grand Island plant people, who had been building red combines since 2004. The event was attended by CNH executives, including our CEO. The event was particularly special to those of us who had been with the Axial-Flow since the beginning. We got a chance to reflect on the time as well as connect with old friends.

Some of the people who created the original Axial-Flow combine got up and spoke for a bit, sitting on hay bales set up for the event. We also had a few longtime customers speak.

Chet Eyer, who had purchased one of the first 300 Axial-Flows in 1977, spoke for a bit. During his talk, he told me he had something for me, and took off his belt. He had the commemorative belt buckle given to him in '77.

I told him to keep it and give it to his grandkids.

Chet's gesture meant a lot. That day, we celebrated more than just innovations—we honored three decades of working together, all while bringing a new factory and corporate partners into the fold.

1977 Original 300 Belt Buckle

▲ These commemorative belt buckles were given to the 300 customers who purchased Axial-Flow combines in 1977. *Lee Klancher*

30 Years of the Axial-Flow

▲ In 2007, the Axial-Flow combine was 30 years old. Case IH celebrated the anniversary with special badges and events. *Case IH*

THE 100 CLUB

By Gerry Salzman

In 2007, John Arnold noticed something interesting. Rob and Sue Holland, his longtime customers who owned a custom-cutting outfit, were about to purchase their one hundredth Axial-Flow combine. John called and told me about it, and I thought that was pretty remarkable and well worth celebrating.

After doing a bit of research with other Case IH dealers, we discovered that four other customers had purchased one hundred or more Axial-Flow combines. All were honored at a ceremony held at the Grand Island, Nebraska, plant, where the Hollands were able to watch their one hundredth Axial-Flow roll off the line.

Steve Lee, the retired Grand Island plant manager, rolled out the red carpet for them. "The day we brought them to the plant, I actually arranged to shut down the entire plant, and all the employees lined the route that we were taking to our visitor's center and gave the customers a standing ovation that just knocked their socks off," he said. "They were really pleased with that. Each of them received a banner from the plant that was signed by every employee."

Rick Farris described the event as a bit of a reunion. "Oh, that was great," he said. "There were several people there, virtually all of them I'd met at some time or other prior to that 100 Club deal. A lot of the people that came from the Case IH side I knew already too. We've worked pretty hard with the engineering groups, especially the Mount Joy group over the years on the machines."

The Hollands were particularly touched by the experience.

"It was very overwhelming for both of us, to be honored and appreciated like that," Rob Holland said. "When they opened the doors to the plant and all the employees were standing there, clapping and thanking us, it was really emotional. We've just never experienced anything like that, and we were feeling gratitude to the employees who build the quality we need to help us do what we do."

100 CLUB LIST

As of 2015, the following people had purchased 100 or more Axial-Flow combines since 1977. The year they were put in the club is noted in parentheses.

Doug Paxton (2007)
Paxton Harvesting
Weatherford, Oklahoma

Rick & Pat Farris (2007)
Farris Harvesting
Edson, Kansas

Rob & Sue Holland (2007)
Holland Harvesting Inc.
Litchfield, Minnesota

Mark and Ken Klatt (2007)
Klatt Harvesting
Foremost, Alberta

Gene & Shirley Linnebur (2007)
Linnebur Grain & Buffalo Farm
Byers, Colorado

Keller Brothers Harvesting (2010)
Duane and Cindy Keller
Jerry and Dale Keller
Ellis, Kansas

Johnson Harvesting (2012)
Lance and Shawn Johnson
Evansville, Minnesota

Douglas Inglis, Farms Director (2013)
Velcourt Ltd.
Suffolk, England

Ben and Gerald Walters (2014)
Walters Harvesting
Lethbridge, Alberta, Canada

CASE IH
2588
THE NORTH FACE

Chapter Eight

2009–2015 THE MODERN ERA

By Lee Klancher

"I believe the major industry strength of Case IH is being close to customers, and understanding their needs drives the company's innovations."
—Andreas Klauser, Brand President, Case IH Agriculture

In 1977, Chet Eyer and his son, Dan, purchased one of the original 300 Axial-Flow combines. The farmers from Illinois were originally attracted to the machine because it provided a cleaner sample.

Today, they farm about 3,700 acres of ground, with 2,200 of that in corn and 1,500 acres of beans. Their primary combine is a 7120.

"All of our beans are seed for two major seed companies and we get paid on quality from those seed companies by the number of splits and everything, and we've just had excellent clean-outs with our combines," Dan said. "So that's one reason we stay with our red combines."

Another is the relationship they enjoy with Case IH. The people Chet met with in 1977 to set up that original Axial-Flow are still close to the Eyers. Those close relationships have been part of the combine business for decades, and the business continues to revolve around intimate connections. Nearly everyone who touches a red combine is directly connected to the end user.

Eric Shuman grew up in farming country in southeastern Pennsylvania, and studied mechanical engineering and manufacturing management at Penn State. He has worked as a production supervisor and quality manager in CNH manufacturing plants.

When he came to CNH, he was struck by the close-knit community present in the combine business.

"I think, one thing coming from outside of the combine business in 2007, I mean, I will say that the team not only at Grand Island, but how we react or communicate with product development, it's a very unique team; it's a very tight-knit team," Shuman said. "It's almost like a family atmosphere. A lot of open communication between the product development

Good Times

◀ Net farm income rose from $56 billion in 2009 to record-setting $129 billion in 2013. Farmers and dealers worked together to keep up with the high demand created during this time of growth. *Case IH* team, the plant, and the customer and . . . I think that really is the key to our success.

"It's truly a cross-functional team, and it's everybody providing their thoughts to get to the final solution."

When Gerry Salzman retired, Shuman took over the role of directing product marketing worldwide for Case IH, and he saw that one of the major challenges faced by the company in the modern era was to meet the needs of a global marketplace.

"I think the biggest challenge on it, when you look at a global standpoint of the product, is dealing with the regional conditions and how those are very unique country by country and region by region," Shuman said. "Looking at the product globally, the solution that's right for the North American customer may not be right for the European or Brazilian customer. Those are certainly some of the challenges that we face especially as a global corporation responsible for the quality and performance of a product that we thought was very robust and durable in one condition in North America . . . maybe we face some challenges in Brazil. We had to make some changes to modify or meet the needs of that local market."

Back on the farm, Chet doesn't have to worry about engineering for a global market. He just focuses on what the new technology can do for his and his son's operations. He sees the future as just as interesting now as it was back in 1977.

Unified Design

▼ When the 120 Series and 88 Series were launched, both featured a new body style as well as technical upgrades.

Case IH

"When I started, we could take two rows and you was lucky to drive two miles an hour," Chet said. "Now look what we're doing. So we've a lot to learn yet. We're just beginning."

The modern era of Case IH combines brought to light a complete upgrade to the midrange line of combines and updates to the emissions systems. The line would continue to refine and grow, with more powerful machines able to harvest more quickly and efficiently than ever before.

"Look at these combines we have now," Chet said. "They've got six, eight, ten, twelve rows, and they got combines that can run four, five miles an hour and harvest the corn. And you just think about this."

Innovation will become more and more important, as the world continues to grow and fewer and fewer people will have to feed a larger world. New technological breakthroughs similar to those developed by the original founders of the Axial-Flow will be required to succeed.

"I think what was done with the original Axial-Flow combine epitomized what we are about as a company: technology and innovation," Case IH Vice President Jim Walker said. "We took that risk, and we have to take more risks like that to truly innovate."

Case IH 7120

▼ The model 7120 Axial-Flow replaced the 7010 Axial-Flow in 2009 and is powered by an 8.7-liter Iveco engine rated for 360 hp. This one is harvesting soybeans and is owned by Daniel F. Tordai. *Lee Klancher*

120 SERIES

By Ken Updike

The year 2009 marked the arrival of the 120 Series flagship combines. Case IH had an all-new Axial-Flow family that now included six basic models. The new 120 Series included the new 360-hp 8.7-liter Iveco-powered Model 7120 (Class 7 machine) with a 315-bushel grain tank; the 10.3-liter Iveco-powered 8120 (Class 8) with a 350-bushel grain tank; and the all-new 9120 (Class 9) powered by a 12.9-liter Iveco engine rated at 483 hp and fitted with a 350-bushel grain tank.

The feeder chain on the 120 series was improved from a four-strand/two-slat chain to a four-strand/three-slat design. To pick up the larger headers being offered, an optional heavy-lift package with 90-mm lift cylinders replaced the standard 70-mm header lift cylinders.

For field conditions less than ideal at harvest time, factory- or field-fitted Quadtrac-style rubber drive tracks could replace

rubber tires on the 8120 and 9120. These tracks offer outstanding flotation in wet harvesting conditions that can cause wheeled combines to become stuck. The tracks' low ground pressure also reduces compaction—especially helpful as the size (and weight) of combines continue to grow.

The all-new 126-blade MagnaCut Chopper was also introduced with the 120 series. This chopper cut crop residue finer with less power than the previous design. In 2012 it was revised again as the 40-blade MagnaCut Fine Chopper.

The 120 Series included a new rear deck and service ladder to improve access to the rear of the combine and the engine area, not to mention an optional heated red-leather seat (like that used in the Magnum tractor).

Many combine owners use compressed air daily (or more frequently) to blow off their machines to reduce chaff buildup and the chance of fire. In 2010, an optional engine-driven air compressor was offered as an onboard source of compressed air for all Axial-Flow combines. Air-cleaning is preferred over

Case IH 7120

▼ The rear-opening service panels on the 7120 are made of a composite material that won't dent or rust. These hinged panels allow access for cleaning and servicing. *Case IH*

Case IH 8120

▲ The 8120 replaced the 8010 in 2009. This cutaway view shows the inner workings of the 8120 combine, including the AFX rotor and cross-flow cleaning fan. *Case IH* water or other methods because no residue is left to collect dirt or dust. Many times, too, water allows dirt to accumulate in hidden or hard-to-reach areas where it becomes packed in and causes rust.

Case IH has adjusted the unloading auger design several times since the combine's 2009 introduction. The unloading auger discharge boot has undergone the most revisions. The most recent innovation has been the development of a powered auger discharge boot. Here, an electric screw motor (linear accelerator) is connected to a pivoting spout controlled by a switch in the cab. Moving the spout allows the combine to travel in a straight line when unloading and still place the grain in the far side or corner of the grain cart or truck. This feature is very important for those who unload on the go.

Case IH 9120

▲ The 9120 Axial-Flow has a 12.9-liter Iveco engine rated at 483 hp. The 126-knife MagnaCut chopper was first introduced on this model. This 9120 has the optional heated red leather seat too! *Case IH*

88 SERIES

These midrange combines are highly evolved versions of the original Axial-Flow, a testament to the capability of that platform. Where the original model was the class leader, this midrange series now provides powerful, capable mid-sized machines offered at more affordable prices.

These new 88 series machines debuted in 2009. The series included the 5088 with its 8.3-liter engine and 250-bushel grain bin, which replaced the Model 2577; the 6088 with an 8.3-liter engine rated at 305 hp and a 300-bushel grain tank, which replaced the 2588; and the 7088 with its 9.0-liter engine and 300-bushel grain tank. All three models had the front axle moved 4 inches ahead and the rear axle moved 5 inches back. The longer wheelbase gave a smoother ride and better load-carrying capabilities. Both the 6088 and the 7088 use a rotor-drive belt that is 0.75 inch wider than the 2588's.

Case IH introduced several new unloading auger changes to the Axial-Flow family in 2009. Until 2006, the discharge end of the unloading auger was a rubber straight down boot commonly called a bottom discharge. In 2007 an angled plastic boot was introduced that discharged the grains from the end of the unloader tube. Case IH changed this design in

88 Series

▼ This image shows the entire line of 88 series combines for the 2009 season: the 5088, 6088, and 7088.

Case IH

June 2009 to maximize grain savings and reduce overall wear at the end of the auger by using a new bottom-hinged grain-saver door and a compatible unloader spout. To accomplish this, the unloader auger tube was shortened by 15 inches. In July 2009 the discharge angle of the auger was updated again to minimize the risk of trapping grain inside. A field improvement campaign was instigated to update the 2007 and 2008 model year machines. Current Axial-Flows use this "new" unloading auger setup.

Inside the cab is a common propulsion handle like that used on the 120 Series. At the rear of the combine, there's a stationary radiator air screen with a hydraulically driven revolving black plastic vacuum wand to keep debris off the screen.

In 2012 the grain tank on the 5088 was upgraded to be optionally available at 300-bushel capacity, and a chaff spreader was introduced at the rear of the combine under the larger straw spreaders to spread the fine chaff that exits the combine from the lower sieve—especially useful in soybean crops.

Case IH 7088 with Processor-Collector

▲ Testing combines in the real world is vital to their success—the conditions combines operate in are almost infinitely variable. This setup allows test engineers to collect extensive data on combine performance. *Case IH*

Advanced Farming Systems

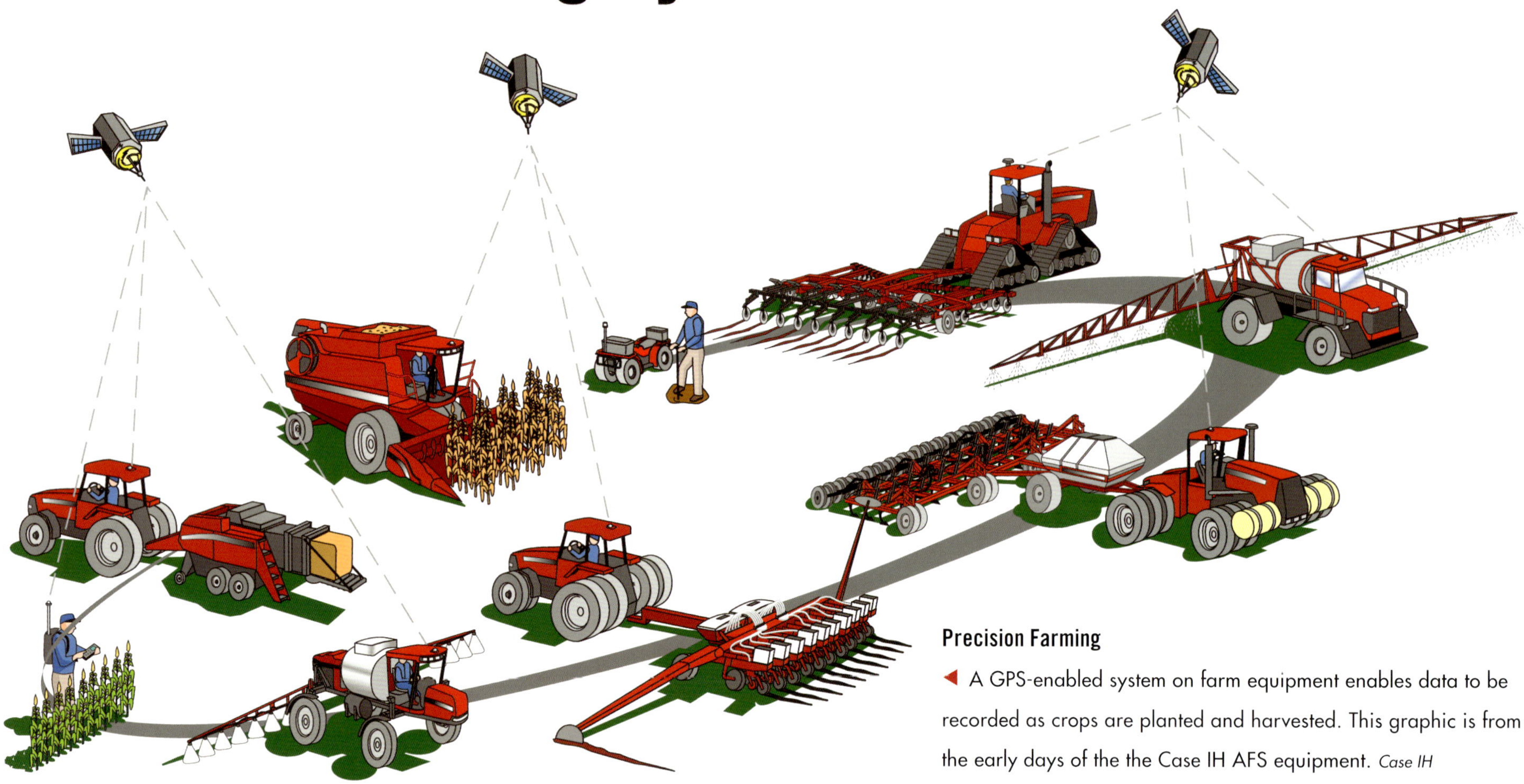

Precision Farming

◀ A GPS-enabled system on farm equipment enables data to be recorded as crops are planted and harvested. This graphic is from the early days of the the Case IH AFS equipment. *Case IH*

THE EVOLUTION OF AFS

By Lee Klancher

In 1983, Korean Air Lines Flight 007 strayed into the USSR's prohibited airspace. The Boeing 747 was shot down and all 269 people on board were killed. As a result, President Ronald Reagan decreed that the Global Positioning System (GPS) that was under development by the U.S. Department of Defense be made freely available to the public.

The first GPS satellite was launched in 1989. By 1995, with twenty-four GPS satellites in position, the GPS "constellation" was complete enough that GPS data could be used for accurate geo-location—to a point. The signal available to civilians was still limited in its accuracy.

As those satellites came on, a flurry of development occurred as innovators looked to use GPS technology for farming. GPS allowed farmers to link fertilizer-application rates, seed-planting rates, and other map-based "prescriptions" with GPS units installed in machines in the field to apply the appropriate amount of nutrients and seeds to each spot in the field.

But this technology can be used to do more than just spread fertilizer: a computer in the tractor or combine cab can map crop yields and make adjustments during planting,

Data Access

◀ AFS has progressed dramatically since it was first introduced in 1996. With AFS Connect, farmers can access data about their machine and crops from their desktop computer or any other device.

Case IH

fertilization, or harvest. It also allows the ability for auto-guidance, with the GPS system keeping the rows perfectly straight and spaced.

This site-specific crop management became known as precision farming. In 1995, Case IH started building a team to launch its precision farming offering: Advanced Farming Systems (AFS).

David Larson's and Steve Faivre's company, AgriCAD Inc., was one of several firms working in the early 1990s to figure out how to use GPS data on the farm. Technology developed by their company was purchased by Case IH under the guidance of Jim Stoddart, who worked closely with Larson and Faivre, along with Linda Knoll and a small team, to create AFS. Larson eventually went to work for the company, first as the general manager of AFS Services and later as a vice president of product portfolio management. Knoll took over leadership of AFS from Stoddart. He later filled a series of leadership roles in the company, and Knoll would eventually become the CNH vice president of human resources.

According to Larson, one key driver for the use of GPS was the development of crop yield monitors in the early 1990s. Yield monitors measured how much crop was being harvested from the field by measuring the bushels of grain coming into the combine bin along with the moisture of the grain. The yield monitor, when tied to GPS, could create a map of the yield for the harvested area.

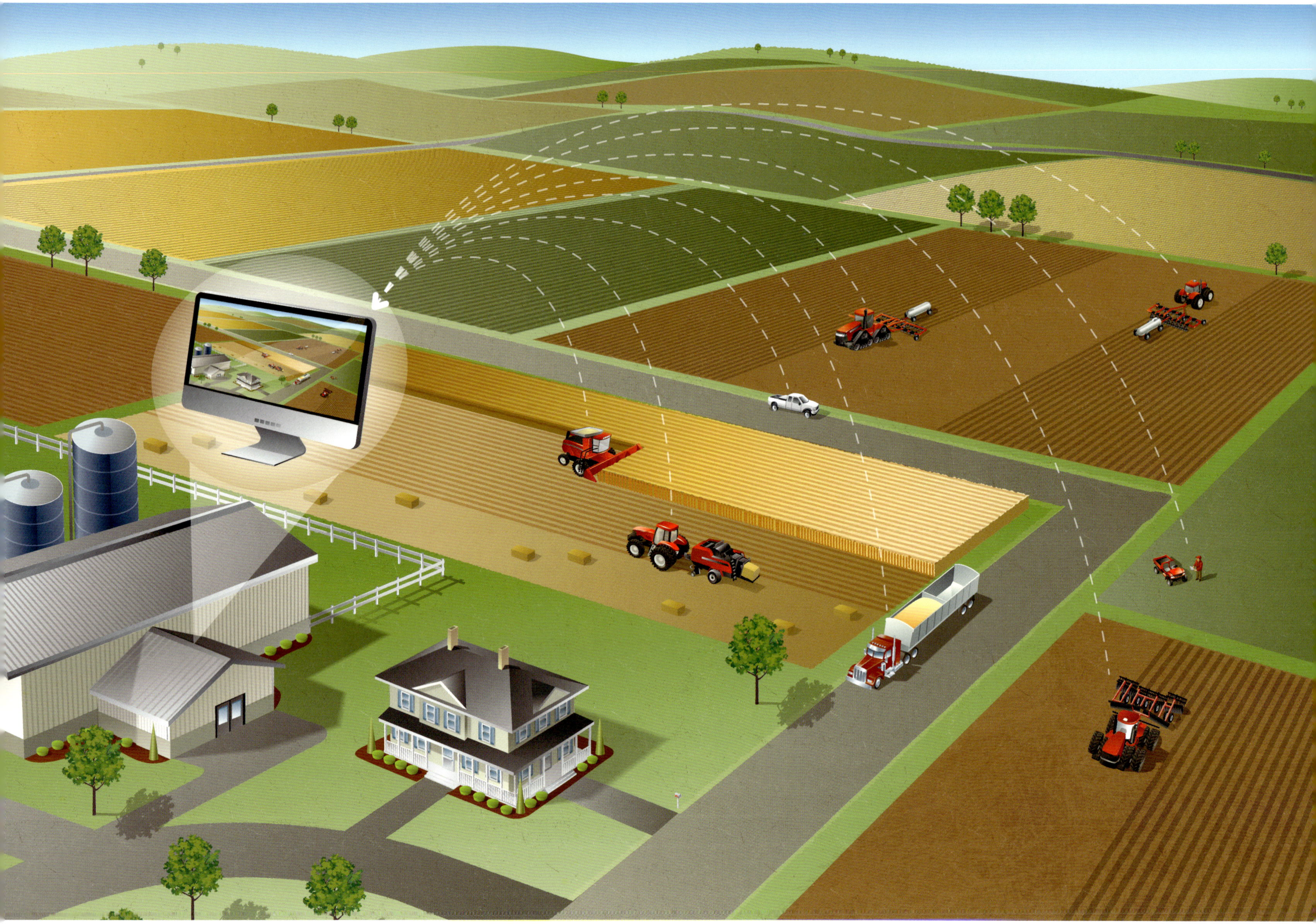

The Connected Farm

▲ Data transfer has changed dramatically with AFS systems. As of 2015, all the planting, harvesting, and fertilizing data could be sent to a central computer via a cellular phone network. *Case IH*

AFS began from a simple idea. "Our objective was to develop machines that could collect and utilize agronomic-based data to create and execute a plan in the field," Larson said.

When he came on, James Stoddart was leading the AFS group and it consisted of eight people. The group first had an operating budget in early 1996. That year, they released their yield monitor and AFS began to quickly ramp up.

By the end of 1997, they had approximately sixty people working in the group and had a system that could control application of seeds and fertilizer. The technology allowed much to be accomplished, which generated even more interest in AFS. The AFS team paired with Trimble and Ag Leader Technology to develop key components of the system. Trimble brought class-leading precision GPS data to the table, while Ag Leader Technology had the best yield monitors.

Not long after launching, the AFS group privately demonstrated a tractor that could operate autonomously using GPS guidance integrated with machine-control systems. The technology promised great things and the AFS group continued to grow.

Precise positioning was an issue with the early GPS systems due to intentional degradation of the GPS accuracy by the U.S. Department of Defense (known as selective availability). Errors were also created by the impact of the atmosphere on the signals from GPS satellites. The way to overcome the error in position is with a technology called Real-Time Kinematics (RTK). RTK utilizes fixed position "base stations" whose locations are precisely mapped. Since a base station's precise position is known, RTK can compute the error in the position information coming from GPS satellites. The base stations are linked to nearby machines operating in the field via radio technology to continuously send positioning corrections.

In the early days, customers could purchase base stations, but they cost about $40,000. In 2000, the deliberate degradation of the GPS signal's accuracy was eliminated, which improved the accuracy of the system. The network has continued to improve and now is accurate to within a few feet. For those who need more precise feeds, subscriptions are available for feeds that use RTK base station networks.

In 2011, the addition of telematics via AFS Connect allowed AFS to record data from any piece of equipment on the farm using the GSM (cell phone) network or a mobile radio network. This means that data from multiple tractors and combines can be captured and consolidated. The data captured includes fuel consumption, crop yields, moisture content, machine settings, and alarm or maintenance messages.

The automatic steering system has become quite sophisticated and can correct for tilting and jolting in the field, as well as adapting to contours in the terrain.

At some point, this will allow equipment operation to be entirely automated. The giant trucks used in open-pit mines are already fully automated, but the environment has controlled access and more stable conditions than a field.

According to Larson, many hurdles have to be cleared before the technology can be applied on the farm.

"In places like western Australia, there is nobody there and the field rows are miles long," he explained. "There is no reason you couldn't have a setup where an operator runs one machine and another runs close by autonomously." The second machine could follow the path of the first, and the operator could shut it down in case of any emergency.

In the meantime, AFS on 2015 machinery allows auto-guidance to precisely plant rows, and it gathers extensive data that allows the farmer to maximize yield.

While auto-guidance is fun to think about, the immediate future might be determined by how data is effectively managed. Jim Walker, the vice president of Case IH Agriculture, said that data integration implemented by AFS is perhaps the most significant new technology on the farm today.

"I would say that data management and data utilization is the new iron of the future," Walker said.

130 AND 230 SERIES

By Ken Updike

The announcement of the new 130 Series midrange Axial-Flow combines in 2012 marked an important milestone in Case IH history. With the new emissions standards, Case IH moved engine production from the Consolidated Diesel Company (CDC) to FPT Industrial. This turned out to be the perfect time to make the move. FPT was part of the entire organization, so it made sense to use its engines. When the new emission regulations came into effect, the time was right to switch. "We were blessed to have FPT's experience with emissions," Gerry Salzman said. "So that worked out really well." The new engines met Tier 4A and 4B standards, and also offered improved efficiency.

Three new models were introduced in the series. The 5130 has a 6.7-liter engine rated at 265 hp and a grain tank capacity of 250/300 bushels. The 5130 replaced the 5088. The

Case IH 7130

▶ This 7130 Axial-Flow has a 2608 eight-row 30-inch-spacing corn header attachment. The 7130 is fitted with the new common cab, as evidenced by the cab's chrome-trimmed headlights. *Case IH*

6130 replaced the 6088 and has an 8.7-liter engine rated at 320 hp and a 300-bushel grain tank. The largest model in the new series is the 7130, with an 8.7-liter engine rated at 350 hp.

In the feeder house of the combine, chain guides were added to the front feeder chain drum to improve feeder chain performance and extend feeder chain life in all crop conditions.

With the use of large (40-foot) draper headers, it became evident that the unloading auger lengths needed to grow. This presented a problem both in Europe and North America. In Europe, adding length to the auger could cause it to strike something while turning in the field or during transport on the roadway. In America, it was an issue with machine storage. Since the auger was extended so far beyond the rear of the machine, it was difficult to park inside a shed without having the auger strike something.

A folding auger was devised as a remedy. Case IH experimented and patented several folding unloader augers. The design that was sent to production has the rear section of the auger fold at a 90-degree angle to the machine to tuck behind it. This reduces the overall length of the machine and does not increase its width. The folding unloader auger is optional equipment. To provide additional safety, the unloading auger operation is connected to the operator presence seat switch. If the operator is out of the seat for more than five seconds when the unloading auger is engaged, the auger shuts off automatically.

Another new feature on the 130 Series is the folding grain tank cover option. Here, four hinged panels create a grain tank extension that can fold flat for transport or storage via a cab-operated rocker switch. This option is very useful and popular in both Europe and North America.

The 230 Series flagship models debuted in 2012. The smallest of these three models was the 7230 that replaced the 7120. This has an 8.7-liter FPT engine rated at 380 hp and a 315-bushel grain bin. The midsized model is the 8230 that replaced the 8120. The 8230 has a 12.9-liter FPT engine

Case IH 7230

▲ This 7230 is fitted with optional rubber drive tracks. Its 8.7-liter, 380-hp FPT engine is Tier 4B–rated. This 7230 is marked with 20K decals, indicating it is for the European market. *Case IH*

Case IH 8230

▲ An 8230 fitted with a draper header rests after harvesting wheat. The draper head has two reels to reduce the load in tough crop conditions. *Case IH*

Case IH 9230

▲ Unloading wheat into a grain cart is easier with the pivoting spout option on the 9230. This allows the operator more control of the grain stream when unloading. *Case IH*

rated at 450 hp and a 350-bushel grain bin. The big one is the 9230, which has a 12.9-liter FPT engine rated at 500 hp and a 350-bushel grain bin. The 230 Series combines are Tier 4A emissions compliant.

The grain tank unloading system was completely redesigned to accommodate a new, large unloading auger that rated 4.0 bushels per second on the 7230 and 8230, and 4.45 bushels per second on the 9230. The unloading auger was increased from 12.6 inches to 14 inches in diameter and was offered in longer lengths to accommodate larger 45-foot-wide headers.

There are two continuously variable transmission (CVT) drive systems on the 230 Series Axial-Flow combines. One CVT drives the feeder and header jackshaft; the other drives the rotor. These provide power and control for the feeder and the rotor. Using a very simple and efficient design, the CVT system allows the operator to vary the speed of either the feeder or the rotor.

The 230 Series combines have a cross-flow fan designed to provide even airflow to the sieves. Airflow is pulled from the entire top of the fan housing. (It does not pull air just from the sides like a paddle fan does.) This provides even airflow across the entire width of the sieves. The cleaning fan is hydraulically driven and may be adjusted electrically from the cab between 300 and 1,150 rpm. The cross-flow fan uses 40 small blades (20 per half) held by composite discs in five locations. The blades are slightly advanced in the middle (i.e., they are not straight across). Each blade can be removed separately if needed. The fan operates in a closed-loop control system, meaning that the fan speed will be maintained at the desired level anytime the engine rpm is above 1,800; when the engine drops below 1,800 rpm, the fan runs at its maximum speed. If the engine is pulled down, the fan speed can be maintained at its maximum speed down to approximately 1,900 engine rpm.

Working with the cross-flow fan are the self-leveling cleaning sieves. The entire cleaning system self-levels up to 12 percent in each direction. The active grain pan and cleaning fan are also self-leveling. Cleaning size really does matter. Case IH does not disappoint with its industry-leading cleaning systems. Flagship combines (7230–9230) set the standard with a whopping 10,075-square-inch sieve area.

Any crop material that falls through the last few rows of the upper sieve or off the rear of the lower sieve is directed to the tailings auger and the tailings processor. The only material that should be found in the tailings is a very small number of unthreshed grains. The processor will rethresh these grains and deliver them back to the upper sieve for cleaning.

The tailings processor uses a series of three four-bladed impellers to transport the material. As the material is discharged from the tailings auger, it enters the lower impeller unit. The material is pressed between the impeller paddles and the clean-out door that works as a concave. The aggressive movement of material across the paddle and concave should perform the required rethreshing. The Tri-Sweep Processors use a 21-inch center impeller to reduce the distance between the lower and center impellers. This provides better dirt and/or wet-material handling. The center impeller's tip speed is approximately 5 percent faster than the lower impeller, and the upper impeller's tip speed is approximately 10 percent faster than the center impeller.

After being threshed, the grain is elevated to the grain tank where the unloading systems on the 7230, 8230, and 9230 have been improved to include higher unload rates, longer unload augers, and the ability to control the horizontal auger independently (dual drive). The 7230 and 8230 unload at 4.0 bushels per second, while the 9230 unloads at 4.5 bushels per second. The 9230 comes standard with the dual-drive unload system. The entire system has been redesigned, not just sped up or lengthened, to handle the new lengths and rates to promote increased operating efficiency, grain quality, and durability.

Case IH 7130 Controls

▼ An interior cab view showing the multifunction control handle and the touch-screen display, which replaces the A-post gauge cluster used in prior models. There is even a storage space for a cellphone by the touch-screen. *Case IH*

BUILDING RED COMBINES

Everyone who works at the combine plant in Grand Island, Nebraska, has driven a combine. Not just the people who work on the floor building the machines, but everyone from the person who greets you at the front door to the night watchman who makes sure you don't sneak in late at night.

Steve Lee, the retired longtime plant manager at Grand Island, explained why he believes that extra step was necessary.

"I told people many times that other than buying more land to farm, the combine is probably the most significant investment our customers will ever make," he said. "To help promote that sensitivity we had everyone from human resources, accountants, as well as people on the line, out driving combines."

All North American, European, and Australian Case IH combines are built at the plant in Grand Island, Nebraska. In addition, Grand Island also serves selected markets in South America and other international regions.

Other Case IH combine manufacturing plants are located in Sorocaba, Brazil; Cordoba, Argentina; and Harbin, China. By the end of 2015, the plant in Curitiba, Brazil, will be a combine header plant only, serving primarily the South American markets.

The Grand Island facility opened in 1965 to conduct final assembly of New Holland combines and added the Case IH Combine line in 2003. Some of the workers there have been building combines for more than 40 years.

Beyond everyone involved having driven a combine, many of the people working the line are, or were, farmers. "We're in the heartland of agriculture, certainly where combines are used so a lot of employees either came off the farm or have relatives associated with agriculture," Lee said.

Bill Baasch is the plant manager at Grand Island, and he said that the plant focuses heavily on using modern management and techniques. This means that when anyone in the plant identifies a potential improvement, the suggestions are sought out and acted upon.

"We continually utilize employee suggestions and work with design engineering to come up with process improvements in how we build our combines," Baasch said. "The goal is to reduce production manufacturing costs while improving product quality."

He also added that interaction with customers is part of the entire experience. People from across the company understand and appreciate the unique link between the customers and people who design, build, and sell combines.

"We really try to focus in on customer satisfaction and the quality of the product," Baasch said. "We've got a very strong workforce out on the shop floor who understands the importance of building quality into the combine."

The Grand Island plant offers tours to customers, some of whom opt to be at the plant when their machine comes off the line. Lee said about 4,000 customers tour the plant each year, and those tours are described by all involved as special events.

"Our 'drive off' customers are the first to crank the engine over in our plant. They are allowed to sit beside our employees as we conduct full validation tests of the product. The customer watches as we assemble the components to the product, and at the end of the day they drive the combine out of the assembly building," Baasch said. "Throughout the entire process, they are meeting the proud employees in our plant. Grand Island conducts approximately 250 customer drive-offs a year. It is these types of customer interfaces that make building combines in Grand Island so special to me."

The plant has modern facilities and processes that can build and finish parts on demand, which reduces the need for excessive inventory and allows the plant to more closely match supply of combines with the demand.

Ultimately, however, the plant's strength is a Midwestern heart and soul.

◀ The metal frame of the combine is welded together. While robotics do much of the work building a combine, plenty is still done by skilled welders. *Lee Klancher*

▶ This machine and operator are performing the fine art of balancing the rotor. Older rotors can be refitted with new rasp bars and then sent to the plant to be rebalanced. *Lee Klancher*

▲ The rotor tube is being welded together. *Lee Klancher*

▶ The painted and assembled lower units travel down the production line. The combine will stop at 31 stations in order to get through the assembly line. *Lee Klancher*

◀ These build sheets attached to the frame describe how the combine is finished and outfitted. The sheets travel with the frame, and the machine's serial number is already assigned. *Lee Klancher*

▶ The rotor and housing slide into the lower assembly. *Lee Klancher*

◀ Here, the top portion of the combine drops onto the lower assembly. *Lee Klancher*

▶ The cab is mated to the combine. *Lee Klancher*

▶ Six of the 31 stations on the assembly line are dedicated to testing. This is one of the test stations. *Lee Klancher*

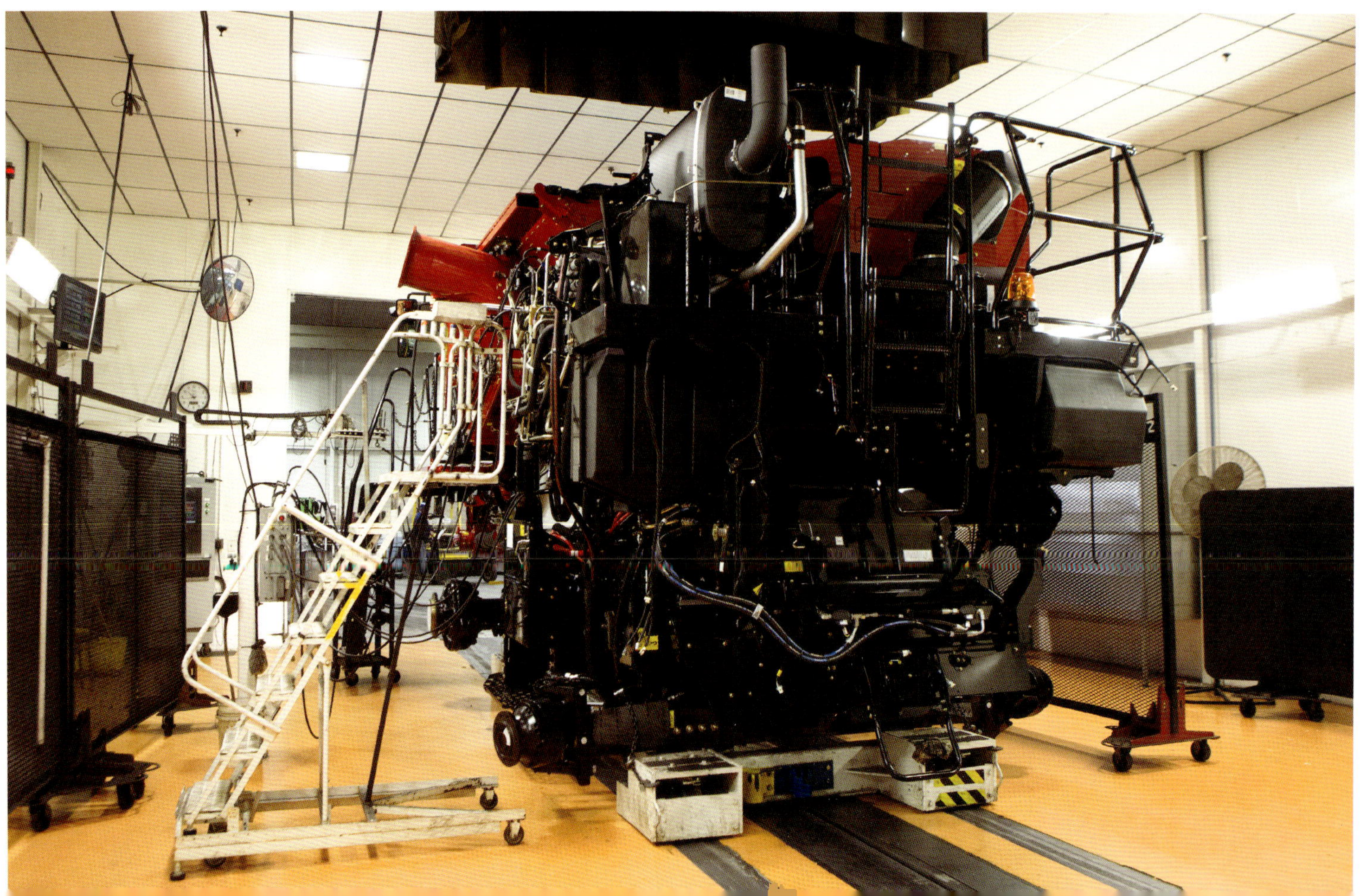

▲ A row of red combines, ready to ship. Case IH combines are shipped all over the world. They travel via truck to North American destinations. Overseas combines travel by rail to the port in Baltimore and then head abroad on cargo ships. *Lee Klancher*

140 AND 240 SERIES

By Ken Updike

The new 140 Series midrange combines that debuted in September 2013 for the 2014 model year comprise the 5140 that replaced the 5130; the 6140 that replaced the 6130; and the 7140 that replaced the 7130. The 140 Series are Tier 4B emissions compliant, and all three models have FPT engines with patented Selective Catalytic Reduction (SCR) emissions technology.

Tier 4B emissions require higher combustion temperatures and cylinder pressures in the engine, so Case IH developed a coolant that resists deposit buildup and other issues that can occur in these extreme conditions. Actifull OT Extended-Life coolant delivers full protection in all engines, including Case IH Tier 4B machines. This new coolant is *not* compatible with any other type of engine coolant (either ethylene or propylene glycol). Any attempt to intermix the new coolant with the others will result in the two precipitating out and possibly gelling inside the engine and radiator.

Case IH 7140

▶ A 7140 Axial-Flow fitted with a the new-for-2014 3162 draper head. The manufacturing plant in Burlington, Iowa, was upgraded to build draper and corn heads. *Case IH*

Starting in 2014, the Actifull OT coolant could be retrofitted in older models by first flushing the coolant system with clean water at least three times to remove any traces of the old coolant. A decal noting that the combine has Actifull OT coolant is applied near the radiator cap/overflow tank to alert the operator or servicer.

In 2013, Case IH debuted a "common cab"—an operator's cab shared by both the Flagship (230) and midrange (130) combines. This cab can be identified by the chrome strip on its headlights (non-common cab models have this strip painted black). Another identifier is inside the cab, where the common cab has the electrical fuses located to the right and behind the operator's seat. The new cab was introduced with September 2012 production; about a quarter of 130 and 140 Series production for the model year was built before the new cab was introduced. The new cab was offered in two configurations: Deluxe and Luxury.

A few features of the new cab include additional storage space, new steering column, new roof and lighting package, repositioned fuse box, improved visibility, new platform and ladder, and a new right-hand console layout and propulsion lever. There was even an optional Bluetooth-equipped radio! The redesigned platform and ladder made setup easier for the dealer and increased operator visibility. The AFS Pro 700 display became standard equipment for the North American market while the A-Post monitor remained an option for all other markets.

Starting in model year 2014, the lateral tilt option became standard equipment.

A change that was made in the rotor cage area to improve threshing in high moisture and green leafy crop material was the option of a new round-bar concave design. Round bar concaves had been available for many years in Europe, and the Case IH team in Brazil developed them for use in the Latin American markets. The North American engineering team made a few modifications to manage performance and those were made available through the aftermarket parts organization for Axial-Flow combines.

The use of LED (light-emitting diode) lighting began in 2014 when the ultra-distance lighting package became available. This is an aftermarket product that can be installed at a Case IH dealership. This package features four adjustable LED lights mounted on the mirror brackets and controlled by a rocker switch in the cab. LED lighting is quickly replacing halogen and HID (high-intensity discharge) lighting. As LED technology advances, the cost of these units continues to decline, making them more affordable for many applications. LED lights draw very little amperage and do not require

Case IH 6140

▲ This 6140 Axial-Flow is harvesting wheat with a rigid draper header 2030. *Case IH*

Case IH 9240

▲ This 9240 is carrying the new-for-2014 4412 folding corn head. The head folds for road transport and includes heavy-duty drives engineered for high-speed harvest and high-yield hybrid crops. The head is manufactured in a newly upgraded facility in Burlington, Iowa. *Case IH*

Case IH 7240

◀ This 7240 is equipped with a 4412 folding corn head. The head has larger stalk rolls and longer stalk roll knives than older versions, meaning they work faster and more efficiently. They also can be set up more quickly for road transport. *Case IH*

expensive ballasts, transformers, or other related components that HID lighting does.

LED lights produce a crisp, bright white light—the closest to real daylight of any artificial light sources. Plus, the LED beam patterns cover more area than traditional lights, and the solid-state construction eliminates fragile bulbs, filaments, and electrodes. This is ideal for the harsh conditions combines face. And since LED lights far outlast incandescent, halogen, and HID lights, they are the best available lighting investment.

Many operators retrofit their Axial-Flows with LED lighting, which can greatly reduce the electrical loads imposed on the combines' electrical system while providing brighter lighting. One drawback to LED lighting, however, is that because it is a "cold light," it does not create enough heat to melt any ice or snow that may accumulate on the lenses. But this is a minor annoyance for the added performance LEDs deliver.

Axial-Flow combines were Tier 4B emission compliant in 2014. This rating was set by the U.S. Environmental Protection Agency (EPA) and the Canadian Environmental Protection Agency (CEPA) to reduce air pollution from on- and off-road vehicles by 2014. The process began in 1996 and comprised four numbered steps—Tier 1 took effect in 1996, Tier 2 in 2002, Tier 3 in 2006, Tier 4a in 2011, and Tier 4b in 2014.

Case IH met this mandate with the use of SCR technology. Basically, SCR treats engine exhaust *after* it has the left the engine by injecting a colorless, harmless urea solution into the exhaust stream that helps significantly lower the levels of particulate matter (PM) and nitrous oxides (NOx) created by

Case IH 8240

▶ Harvesting windrowed wheat with an 8240 fitted with a 3216 windrow pickup head. The dual-drive tires offer more flotation and load-carrying capacity than single wheels. *Case IH*

the engine. Another method used to lower PM and NOx is to recirculate a portion of the exhaust back into the engine to be reburned. This method, called EGR (Exhaust Gas Recirculation), is very taxing on the engine's cooling system. EGR also uses a particulate filter to capture pollutants. But before this can happen, the exhaust must be cooled or the engine's pistons and valves will melt from excessive heat.

EGR requires short engine oil change intervals due to the system's impact on the engine's cooling system. This is seen particularly with large engines under high-duty cycles (like tractors). The particulate filter captures contaminants until a sensor indicates that it is full and needs to be emptied. To do this, the tractor goes into a generation cycle in which the engine automatically detunes itself and raw diesel fuel is injected into the particulate filter. To burn the diesel fuel that then burns out the particulates, the exhaust temperature during regeneration can reach 1,292 degrees Fahrenheit (nearly twice the normal exhaust temperature). It's a bit like a self-cleaning oven. Eventually, the particulate filter needs to be removed and cleaned by trained dealer personnel. Failure to do so can cause damage to the system and filter and may require the filter to be replaced at a cost of nearly $10,000.

SCR uses a much simpler and safer method involving DEF (diesel exhaust fluid)—a nontoxic solution of 67.5 percent purified water and 32.5 percent pure automotive-grade urea. DEF is not a fuel or fuel additive, but a stable and colorless solution that meets ISO standards for purity and composition and is American Petroleum Institute (API) certified.

In an SCR system, DEF is injected into the exhaust stream through a controlled dosing module that varies the application rate. The DEF vaporizes the nitrogen oxide emissions and converts them into harmless nitrogen gas and water vapor (both are natural components of the air we breathe). By injecting the DEF downstream in the exhaust system, Case IH was able to tune the engine for maximum power and performance and allow an average of 10 percent savings in operation cost via enhanced fuel economy.

DEF is stored in a separate, dedicated tank next to the fuel tank. The DEF tank's capacity is about 41 gallons, but it is rated to be filled at every second fuel fill. The DEF tank has a special-sized fill opening that only a DEF nozzle will fit. This ensures that only DEF is pumped into the tank. The DEF tank has a blue fill cap to match blue fill nozzles.

DEF is nontoxic, nonflammable, and nonpolluting. DEF has about the same alkalinity (pH) as baking soda, so it is slightly corrosive. It is recommended that DEF be stored in clean, dedicated plastic tanks. To maximize shelf life, the ideal storage temperature is between 12 and 86 degrees Fahrenheit.

FPT 12.9-Liter Engine

◀ Dealing with new emission standards required massive investments into engine development. The FPT engines that meet the 2015 standards use a Selective Catalytic Reduction (SCR) system, which is more efficient than other methods on the market.

Case IH

Freezing and thawing do not change DEF's chemical properties; if the fluid freezes, an internal tank heater that uses the engine's warmed coolant will thaw the DEF.

Axial-Flow combines have a DEF-level gauge in the A-post instrument cluster. A series of warnings alerts the operator when the DEF level reaches 10 percent or less; if the tank reaches 5 percent capacity, the engine will de-rate, but power will still be available to allow the combine to return to a convenient location for a refill. Since Axial-Flow combines are designed to have their DEF tank filled at every other fuel tank refill, the level can be managed easily.

The consumption rate of DEF varies depending on engine load. In a typical agricultural application, it's 3 to 5 percent. It's estimated that for every gallon of DEF used, 2 gallons of diesel fuel are saved.

240 SERIES: 2015 UPDATE

The most current family of flagship Axial-Flow combines is the 240 Series. These combines are fully Tier 4B emissions compliant and feature the common cab. The 7240 has an 11.1-liter FPT engine rated at 403 hp and a 315-bushel grain bin; the 8240 has a 12.9-liter FPT turbo diesel rated at 480 hp with a 410-bushel grain bin; and the 9240 has a 15.9-liter FPT turbo diesel rated at 550 hp with a 410-bushel grain bin capacity. All three models are more powerful than the previous series courtesy of their larger-displacement engines.

Case IH 9240

▼ A 9240 harvests wheat with a 3162 Terraflex draper head.

Case IH

The wider headers used today require that lateral tilt be fitted on combines for them to be effective harvesting tools. Hence, a lateral feeder tilt feature is standard on the 240 series. To accompany this feature, the feeder house is reinforced in a number of areas to offer greater durability for the added stresses that these headers impose on the combine.

The larger grain tank capacities on the 8240 and 9240 (increased from 350 bushels to an impressive 410), combined with up to 4.5 bushels-per-second unloading rates, means these combines have the capacity to open up the field while being able to unload in a short distance.

To match the new, larger engines, the fuel tank capacities were increased on all 240 Series combines to allow for a full day's harvest without refilling. This allows operators to start early in the morning and run into the night without stopping for fuel. The 7240 and 8240 utilize the same single fuel tank design. Because of its larger 15.9-liter engine, the 9240 utilizes two fuel tanks. Two fuel-fill points are easily accessed from the operator's platform. While harvesting, the combine pulls fuel evenly from both tanks.

The biggest change to the 240 Series cooling system was the addition of a stationary air screen that provides many advantages over a rotary screen. To keep the screen clean, a wand spins and evacuates debris from the screen. This ensures plenty of airflow to the coolers when harvesting in high-debris areas. The design is similar to that used on the 140 Series.

The 9240 was given a new hydraulically driven cooling fan. At ambient air temperatures—100 degrees Fahrenheit and below—this allows the cooling fan speed to be driven more slowly than if the fan was driven by a fixed mechanical system. The slower speeds require less power, which allows more engine power to be applied to the threshing and header if needed. At higher ambient temperatures—greater than 100–105 degrees Fahrenheit—the fan must turn faster to provide cooling. The 7240 and 8240 continued to use a belt-driven cooling fan.

The 240 Series' electrical system has been improved in a few key areas for better reliability. An electrical controller called an H-Bridge is located on the right front side of the combine below the cab. This works as a relay and is required to control certain electrical options. The H-Bridge reduces the chances of electrical overloads on the various circuits on the combine.

The choice of either 30- or 36-inch-wide rubber tracks (instead of tires) is available on the 240 Series. These are the same proven tracks used on Steiger Quadtrac. The track undercarriage is the same on both options (only the track-width is different). The 30-inch track provides advantages when harvesting raised beds, and since the track is the same width as the boggy wheels, it prevents the belt from folding up and riding on the edge of the idlers and wheels. Both single- and dual-wheel arrangements are also available.

Case IH 240 Series System

▼ This cutaway view of the Flagship combine threshing system shows the four-strand feeder chain that delivers the crop to the AFX rotor. Beneath the rotor, a cross-flow cleaning fan delivers an air blast to the sieves to clean the grain from the chaff. *Case IH*

CUSTOM HARVESTING

By Lee Klancher

When the Soviet Union invaded Afghanistan in 1979, the Carter administration's response was to embargo the export of grain. The Soviet Union was able to source grain elsewhere, and the effect on American farmers—who were already besieged by falling prices and skyrocketing interest rates—was disastrous.

Rob Holland of Litchfield, Minnesota, had bought himself a nice IH 1480 combine not long before that time as part of his fairly new farm. When the embargo hit, his farm hit rock bottom. He flat ran out of cash.

He had financed the combine through International Harvester, and he couldn't make the payment.

"I didn't have enough money to make a payment to International Harvester at that time, and my dealer personally, out of his own personal money, made my combine payment for me," Holland said. "He took a chance with me. He wanted to keep us in business here."

Rob and his wife, Sue, didn't think farming on the scale of their operation was going to bring in enough money to stay afloat, but the two didn't want to abandon the life. They decided to take a shot at custom harvesting. Custom cutters, as they are called, travel across the heart of America with a fleet of combines and trucks, harvesting for farms that don't have their own combines.

The young couple had to start off small. They owned one combine, that 1480, and it would have to suffice. It was all they had.

The Traveling Show

▼ Custom harvesters travel across the country, harvesting for farmers on contract. This is Hoffart Harvesting out of Rugby, North Dakota. Joe and Willie Hoffart used IH combines for decades, even prior to the Axial-Flow. *Case IH*

"My wife and I took off with one combine to Texas because I wanted to stay in the field," Holland said. "We had to stay with agriculture. I like machinery. I like good equipment. We wanted to make this custom harvesting thing work. Sue drove the truck and I was in the combine until we eventually went to two combines, then she ran it. She still goes in a combine if needed for wheat and always in the fall."

They steadily built up a list of customers, traveling from Texas up into Canada each year, following the harvest.

"It took a few years; it took three or four years there to get a route established to where we could see that we could make it work," Holland said. "We eventually grew it into running seven combines for wheat harvest with often more for fall harvest and have continued on with this."

Holland travels with a fleet of workers, semi-trucks, and those seven combines each year. He's one of many custom cutters whose business is contract harvesting for farmers.

The profession is not a new one. Threshers traveled the countryside separating grain since the early 1900s, and small combine operators did the same in the 1920s.

Gene Farris worked on a threshing crew in 1919, when he was 13 years old. He took a job as a mechanic at a local Massey-Harris dealership. At the start of World War II, combines were in short supply and most of them were sold to custom cutters. The young enterprising Gene Farris worked for a year with a custom cutter before he was able to buy two Massey-Harris combines.

Two of his sons, Rick and Gary, continued their father's tradition, and expanded the business by taking fall work

in California, Colorado, Arizona, and New Mexico. The operation, Farris Brothers Inc., became one of the largest in the country.

Rick eventually took over the entire business and stayed with Massey-Harris equipment until the company dissolved. "We looked around, and one of the determining factors was the support from the companies, and Case had by far and again the best support out there," Farris said. "We started purchasing Case combines from 1988, the Case IH brand. I believe since then we've purchased over 160 combines."

Farris favors long relationships and has been working with some of his customers for a long time. "Most of our customers we've worked for a long time—20 to 30 years or more. Where we started, I think this was our 50th year starting there. My job in Nebraska, this is the 49th year for them. The kids are farming it now, the parents are gone, but we've worked mainly for the same families for year after year. Some of the jobs I have at home are people that my father had worked for their grandparents at one time," Farris said. "It's like going home every place you go to work. Our farmers finally get to where they rely on us, but we rely on them too."

Most of the help hired by Farris are college-age kids, often referred to him from colleges with agricultural internship

Seven 2388s

▶ Rob and Sue Holland of Litchfield, Minnesota, run seven Case IH combines in their custom cutting operation. The seven are shown at work west of Regent, North Dakota, in 2005. *Sue Holland*

Wild Harvest

◀ In the 1947 film, *Wild Harvest*, two rival custom harvesters battle for customers and love as they work their way from the Texas panhandle to the Dakotas. Between the fist fights and scheming, the film featured wheat-harvesting teams at work. The film was based on, "The Big Haircut," a screen story by Houston Branch, who wrote 50 screenplays in his career, including the screenplay for *Klondike Kate* and *Mr. Wong, Detective*, starring Boris Karloff.

Public Domain

programs, dealerships, and other people in the industry. While he said that finding help is one of the biggest challenges he sees in his business, he also finds it rewarding to see how some of the kids that work for him move on.

"One of the boys that went to work for me a year ago, right out of high school, went over and went to work at the Deere dealership after we got done there in the end of November," Farris said. "Shoot, once they found out he had all this training and CDLs and all this, they had him delivering machinery and all sorts of things. They've paid his way through technical school now to be a mechanic for them."

Like the Hollands, Farris has built a business that provides harvesting for hundreds of farmers as well as jobs for a wide variety of people.

Back in Minnesota, the gamble Holland's local dealer made by making a payment for Rob and Sue Holland turned out to be a good investment. As of 2015, the Hollands bought 176 combines from their local dealer, Arnold's of Kimball, Minnesota.

Chapter Nine

RED COMBINES INTERNATIONAL

By Lee Klancher

"Farmers today are global agribusiness people. I've always said that what happens around the world will—in one way or another—impact the price of grain."
—Thomas Sleight, President and CEO, U.S. Grains Council

John Machin came to International Harvester in the early 1970s, providing industry training for engineers based in the UK. After five years of the work, his career took a memorable turn.

"About '79, I was called into IH London offices by the marketing director and introduced to something called an Axial-Flow combine," Machin said. "I'd never heard of this damn thing. I looked at pictures of it, I looked at some marketing information and soon found it was a very impressive and very different concept of grain harvesting that had been developed in North America and had established itself very well. It gradually transpired that people tended to see me in the company as Mr. Axial-Flow."

The challenges in Europe at that time were related mainly to weather and residue management. The residue problem was that American farmers of the time didn't care about the straw—they just wanted it out the back. European farmers, on the other hand, relied on quality straw for livestock bedding. The early Axial-Flows were built for American tastes.

The other issue was weather. The first Axial-Flow combines didn't work terribly well in wet conditions. "We had quite a few operational problems because of the conditions in North America—blue skies, low moisture contents, sunshine all the way," Machin said. "When you travel to Europe, you've got rain, rain, and more rain, and it's very wet conditions."

Machin went on to help the Europeans understand and use the Axial-Flow, not to mention passing on feedback back to the United States so the machine's development could be tailored to the market. When the European market began to tail off, Machin found himself facing much more difficult challenges in more exotic locations.

Growing globally was a key part of the strategy of agricultural manufacturers, and a tremendous help to a business that lives and dies by a fluctuating market. As

Harvest the World

◀ Heavy equipment operators are learning to tap global markets in order to help offset the massive investment required to develop agricultural equipment. These combines are at work in Latin America. *Case IH* crop prices go, so go ag equipment sales. Presence in markets around the world makes it more likely that the company will have products being sold somewhere that is having a good strong harvest.

Rick Tolman was at IH in the early 1980s and moved on to become a leader at the Grains Council, which advocates for the grain industry. He would later become CEO of the National Corn Growers. Tolman believes the success of the rotary combine helped drive the interest in business overseas.

"We had satisfied domestic demand," Tolman said. "We were looking for new markets, new opportunities, and then, of course, the Soviet Union was a big market opportunity. Europe had been, but I think in my mind, it's the first time we're really thinking about doing more than just sending our surplus overseas. We were thinking about really servicing those markets, and we had the boom servicing the former Soviet Union for several years and then Europe was winding down.

"The rotary combine connoted more capacity, more larger sizes that the different models came in and being able to go through the field faster in a more efficient way, and new technology. That was also our mindset: 'Hey, we're a productive nation. We're not only going to satisfy our needs, but we're going to provide grain to the rest of the world.'"

In order to provide combines as well as grain to the rest of the world, many challenges would have to be met by people like Machin.

Many of the combines sold in Europe were shipped from America via the ship, the *Atlantic Conveyor*. In 1982, several hundred Axial-Flow combines were supposed to be transported by this ship to Europe.

When Argentina attacked the Falkland Islands, the ship was commissioned by the British government to transport fighter jets. On May 25, 1982, the *Atlantic Conveyor* was hit by an Argentinian Exocet missle and sunk with all hands. Captain North was awarded the Distinquished Service Cross.

"That was a very poignant point for me relating the fact that many of the axial combines were brought over on the ship by a friend of mine, Ian North," Machin said. "He went down with his crew to the Falklands, and the bastards sank him. He went down to the bottom of the South Atlantic."

When Machin went to work in Central Asia in 1996, he saw more problems related to transporting machines. Combines were moved from Helsinki across Russia on slow-moving trains, which delayed delivery schedules and also proved dangerous.

"There were armed guards on the trains because bearing in mind, these trains were only traveling at 30, 40 miles an hour, very slow moving," Machin said. "They would take weeks and weeks to cross the Russian steppe and get to us in Turkmenistan."

Once they did arrive, the trucks in the area were so old and unreliable that operators resorted to driving the combines hundreds of miles to their final destination. Fleets of 20 or 30 combines would be spotted, plodding for hundreds of miles along the highways of Turkmenistan.

Nothing could be taken for granted in these remote parts of the world. Work forces in the area consisted of Soviet trucks loaded with hundreds of school kids pulled away from their studies to harvest cotton.

Combines were damaged, particularly when transported at night and they ran into camels.

Local operators would show up high on a chewable leaf known as "naz" that the Case IH teamed dubbed "love."

Local mechanics attempted to fix electronic systems with their favorite tool—a large hammer.

The challenges Machin faced were unique, but the situation was universal. Selling overseas required solving all the myriad cultural and socioeconomic problems found in each region.

The red combine's journey around the world was not an easy one, but the global marketplace has become critical to the success of Case IH agricultural equipment.

Axial-Flow in Europe

◀ The Axial-Flow combine was exported to Europe in 1979. A few of the Axial-Flow combines were assembled in France in the early 1980s.

Case IH

The global marketplace also has an impact on farmers. While that has been true for decades, it is becoming increasingly important that producers are able to produce for markets in all parts of the world.

Don Fast is a Montana farmer and a past president of the U.S. Grains Council. He has an intimate understanding of the importance of the international market.

"In Montana, we export 80 percent of the grain that we grow. We rely on exports heavily. Half of the wheat crop in the United States is exported," Fast said. He added that one-third of all the agricultural products made in America head overseas.

The quality of the product is paramount for this market, and the rotary combine helped open markets overseas.

"Certain countries complain about quality of grain," Fast said. "With these new innovations with these rotary combines, I think we have a better quality of grain for them. There's less spoilage in shipping and everything."

Fast believes that the key for the future is to embrace the changing technologies. He spent 400 hours in his combine in fall 2014. "Accept change and look forward to the future," he said. "Maybe in 10 years, I can sit someplace else and have the combine run itself."

RED COMBINES IN THE UNITED KINGDOM

By Martin Rickatson

When International Harvester Company founded its British operations in 1906, for many years its business was based on assembling imported equipment shipped in from other IH facilities. The first premises used for this task was in the East End of London and near the offices it had established.

IH Great Britain's (IHGB) first full-scale assembly plant was at Orrell Park, close to the docks in the northwestern city of Liverpool. Subsequently, a much larger works was built in 1938 at Wheatley Hall Road in Doncaster, South Yorkshire. This would become the major manufacturing site for IHGB.

While various companies have built combines in the United Kingdom, IHGB and its successors never manufactured self-propelled harvesters in the UK. But the factory that was to eventually become the main tractor plant did have some harvesting history. At the 1952 Royal Smithfield Show in London's Earls Court, IH launched the International B-64, a trailed harvester thresher with a 6-foot header. The following year the machine went into production at the Doncaster factory.

Driven via a power take-off (PTO) shaft or an optional gasoline engine, the harvester was offered in bagger and tanker versions. It was advertised as being capable of an hourly output in wheat of 2 acres (approximately 3 tons in UK yield terms). Production ended in 1964 after a run of approximately 6,000 units, with Doncaster turning to concentrate on tractor, baler, and construction equipment production. From this point, the British market was served by self-propelled combines made in the company's factory at Lille, France.

UK Harvesting Scene, 1950s

▶ A Doncaster-built B-64 trailed harvester thresher, the only combine ever built in the UK by International Harvester or its successors there, works a British cereal field behind a B-250 tractor in a picture dating from the late 1950s. Before self-propelled combines took over, the B-64 was very successful in the UK, a considerable plus being its straight through threshing system, whereby the 6 foot header fed a 6-foot drum.

Wisconsin Historical Society #117296

Early UK Axial-Flow 1400 Trials

▶ Cereal crops in the UK and across northern Europe tend to be significantly higher-yielding and produce much heavier quantities of damper straw than those in the Axial-Flow's birthplace. While the combines' rotors were up to the job when the first Axial-Flows trialed in the UK, the white-reeled French-made 825 headers like that on this 1440 struggled to handle the densely strawed crop stands and greener materials. *Case IH*

Following the North American launch of the International 1400 Series Axial-Flow combines in 1977, IHGB showed a 1400 Series machine on its stand at the 1978 Royal Smithfield Show in Earls Court. The following summer, five 1460s were brought to the UK. Four were allocated for dealer demonstrations and field trials. The fifth 1460 was put forward by IHGB for "Formal Field Testing" to the UK's National Institute of Agricultural Engineering (NIAE), covering 3,000 acres of cropping across the country. Its researchers found that, when compared with a five-walker IH 953 European-made conventional combine, the 1460 produced significantly lower grain-loss. Lowest losses, at around 1 percent, were achieved with a high rotor speed of 950 rpm and a small concave clearance. At this level the machine's output was around 40 percent greater than that of the 953, even though the 1460 was equipped with a 14-foot header that was relatively small for the size of the machine and performing in heavy-strawed UK conditions.

Roger Arnold, one of the NIAE team that assessed the Axial-Flow, was reported to admit that he was initially skeptical of the principle before becoming involved in its assessment. The tests recorded the time between standing crop entering the combine and straw and chaff leaving the rear as just 2-4 seconds—that compared with 15 seconds for a conventional straw-walker combine. A lower incidence of grain damage was a key benefit.

With the losses staying fairly constant regardless of work rate, it was recommended that the combine was best driven as fast as ground conditions and engine speed would allow, in contrast to the output-versus-losses compromise that governed conventional combine operation. Meanwhile, it was also found that keeping the rotor cage fully loaded gave the cleanest sample.

Early results indicated that the concept had considerable potential in this part of northern Europe, where cereal yields were—and are—among the highest in the world, and crops produce very high quantities of often damp and tough straw.

Header Breakthrough

◀ To overcome the problems experienced with the French-built 825 header in heavy straw conditions, IHGB went on to develop a European-specification alternative, based on the American 810 header to be sold with 1400 and 1600 Series Axial-Flow combines. Modified in-house by IHGB, key attributes of the revised 810 were a greater knife-to-auger distance and hydraulic reel fore/aft adjustment, making more room for the crop before it met the header auger. The European version of the 810 was available in widths up to 20 feet. *Case IH*

Axial-Flows Made in Europe

▶ In an ultimately short-lived venture, International Harvester opened a new factory in Angers, France, in 1982 to build Axial-Flow 1400 series combines for Europe. The 1420, 1440, and 1460 were assembled from components shipped from East Moline, but the 1480 continued to be brought in from the United States. French-made machines were notable for, among other small differences, rear- rather than right-side mounting of their rotary air cleaners to keep within European vehicle-width restrictions. The Angers plant was closed shortly after the sale of the IH ag business to Tenneco. European supply duties reverted to East Moline. *Jean Cointe Collection*

But one of the biggest issues of concern for UK and other European farmers was the effect of the Axial-Flow rotor's threshing action on the straw in drier conditions, where there was a worry the stems could break up. Not only could this result in straw fragments overloading the sieves, but in a part of the world where it is regarded on many farms as a valuable feed and bedding source, it could make the straw that did come out the back harder to bale. It took many years for salesmen to overcome this issue by educating farmers in the best operating techniques for Axial-Flow combines, while the shift from mixed farming to more specialized cropping-only farms—where straw was chopped and thus its quality unimportant—also helped.

Despite a price almost 35 percent higher than that of the 953, then the biggest straw-walker combine in the European IH range, approximately thirty 1460 Axial-Flow models were working across English and Scottish farms in 1980. The company stated that its plan was to have a further 200 units working across Britain by 1981. That figure had to be reined back, though, as strike issues at the East Moline plant reduced factory output and the UK's supply was hit.

Eventually, IH offered the 1420, 1440, 1460, and 1480 models to UK farmers, and the source of machines for the UK was switched for a short time to the new Angers plant in France. When this plant closed as IH hit financial problems, the supply source once again reverted to East Moline.

The American-made headers, though, were found to be unsuited to the heavily strawed crops in the UK and much of the rest of northern Europe. European versions in sizes up to 22.5 feet were developed with greater knife-to-auger distances. Higher straw volumes and transport restrictions caused by narrow roads meant that, at this time, headers wider than this were considered unsuitable for UK-spec machines.

IHGB's Paul Wade was reported in the farming press as saying that drivers he instructed on operating the first 1400 Axial-Flows in the country were initially concerned about plugging the combines and given the advice to drive as fast as conditions would allow. But he noted that this was a rare occurrence and that drivers needed to trust the audible alarm that sounded when rotor speed slowed.

Following the IH ag division purchase by Tenneco at the end of 1984, UK farmers were offered the updated Case IH 1600 Series in 1986, but the 1620 was not brought in. The 1644, 1666, and 1688 models that followed were not offered in the UK, with sufficient numbers of the old series hanging on until the 2100s were ready. Again, the smallest model, the 2144, was judged too small for the UK's tough straw conditions. Likewise, when the 2300s came out, there was no 2344 in the UK offering, while the subsequent 2500 Series comprised just the 2577 and 2588.

The first common platform Axial-Flow UK farmers were able to lay their hands on following the Case IH/New Holland

Royal Seal of Approval

◀ Her Majesty Queen Elizabeth II visits the International Harvester stand at the Royal Show in Stoneleigh, Warwickshire, in 1984. This machine went on to operate on Her Majesty's Sandringham Estate in Norfolk, eastern England. On the Queen's immediate left in the background is John Machin, combine sales manager for IHGB; the man to her immediate right is Mike Croome, IHGB director of agricultural equipment sales and marketing. Also to the Queen's left is Bud Thompson, IHGB managing director, while on his immediate right is Jack Michaels, IHGB director of manufacturing.

Case IH

merger was the 8010, and this was soon topped and tailed by the 7010 and 9010 as UK and other European farmers sought more horsepower and wider headers to boost throughputs in what can be catchy conditions. Since then, through the introduction of the 20, 30, and 40 Series Axial-Flows, the UK offering has largely mirrored that of North America, but with different detail specifications.

John Aaskov joined International Harvester's operation in Copenhagen, Denmark, in 1979 and quickly became deeply involved in the effort to communicate the advantages of the Axial-Flow concept to farmers. But he and those around him were initially unsure as to whether the combines would be offered in Europe.

"We'd heard about them, but at that time there was no Internet or anything, so we really didn't know what they were until we saw them in Denmark and more widely in Europe in 1979," Aaskov recalled. "They represented something of a revolution—people looked at them and said, 'How can a combine with this principle harvest anything?'

"We slowly found out how to run the Axial-Flow and got to know it, and realized that this was a combine which had a great future in front of it."

Aaskov continued, "But farmers in Europe could be extremely traditional and conservative, and it took quite some time to persuade people that this kind of machine could actually work. Germany was particularly challenging due to strong domestic makers, as was Denmark. But some farmers soon realized what an improvement the Axial-Flow offered. Year by year, as the 1400 Series made way for the 1600 Series, we gradually sold more machines in Denmark, and the same happened all over Europe, helped by our field demonstrations. The simplicity of the single-rotor Axial-Flow design, and therefore the low running cost, were key attractions.

Prime Minister Presents Top Trophy

▲ Prime Minister Margaret Thatcher presents the Burke Trophy to Bud Thompson, managing director of IHGB, at the Royal Show, Warwickshire, England, on July 13, 1981. Britain's premier agricultural machinery award was presented to International Harvester in recognition of the innovation behind the design of the Axial-Flow combine and threshing system, and was presented at the country's annual summer farm show. *Wisconsin Historical Society #117303*

"To begin with, sales were slow, but then Axial-Flow sold more strongly for several years. Today it's not the biggest-selling combine in Europe, but farmers here are gradually realizing that the concept is deserving of greater attention."

Aaskov believes much of this is due to European farmers' desire to achieve a good sample in cereals because many of their crops go for quality food production. "Axial-Flow threshing and separation is well respected for producing a good sample due to its gentle action," he explained. "In Europe we also produce a lot of specialist seed crops—vegetable seed, grass seed—and here gentleness is very important too."

Will Bushell began working for IHGB as a combine specialist in 1983 when the difficulties IH was going through were reaching their height. He had worked with Axial-Flow combines, driving them in North America, and knew those conditions well.

"But European conditions are completely different," Bushell pointed out. "I could drive an Axial-Flow, but from a settings point of view, I had to turn to the guys in the UK who had been working with the combine there for the past couple of seasons. They could teach me because they'd already been trying to find the right rotor settings in any given crop conditions."

Just as harvest does, the small team quickly bonded. One of them was John Machin, who really became Mr. Axial-Flow in the UK and even worked as far afield with the machines as Turkmenistan.

"In my first season I was running around pretty much anywhere I was required in the UK, doing demonstrations, helping customers set up their machines, helping dealers with their own demos, fixing, repairing," Machin said. "In those days, you did whatever you needed to do. The system was evolving, right down to working out what Axial-Flow parts to carry on the shelves.

"In my second year, 1984, IH split the UK territory, and I got the south. But during harvest we were still flat out using the machines, helping customers, installing combines with customers. Then through the winter and the spring months, we were helping to close sales. The product knowledge I'd gained and the fact I came from a farm both helped here."

One of the biggest hurdles, Bushell recalled, was gaining the trust of potential customers who had spent months of the year cultivating, planting, and nourishing crops to gain an end product for sale.

"In their eyes we were effectively asking them to risk that—their livelihood—in using a machine completely different to what all the other competitors were selling," Bushell explained. "We had to take combines anywhere, to anyone who would allow us to run them alongside their existing machine and prove them in the field.

"In its earliest guise the Axial-Flow's performance depended on crop conditions. On a hot, sunny day you could easily beat any other combine for performance, but if the sun disappeared or the straw was slightly green, there was less of a performance advantage, although the Axial-Flow would still give a better grain sample.

"That was one of the things that you had to try and get people's heads around—that they needed to operate this machine completely differently. You would begin driving a conventional combine steadily, finding the appropriate settings and speed, then cut the whole of the field at that speed to maintain your sample. With an Axial-Flow, one of the inherent differences is that you pushed it until you ran out of horsepower—it had a tremendous appetite, and it still does today. The more crop you could get in it, the better it performed.

"For many it was a big culture shock. We often found that the best sales we had and the best operators involved people driving the machine who were completely new to combining and hadn't operated a conventional combine before."

New Generation

▲ During the 1985 transition year, as the IH and Case equipment lines and dealer networks were gradually merged, a handful of rare red-roofed Case International 1400 series combines made their way to the UK but imports of the new 1600 series, with a fresh livery, new work-light arrangement, and updated cab interior, were ready in time for the 1986 harvest. This 1660 is on demonstration in Staffordshire, in the west of England, where the hillside conditions are extremely demanding on engine horsepower. Later European Axial-Flow combines received a significant horsepower boost to aid their hill-climbing ability. *Case IH*

2388 X-Clusive

◄ Available only in Europe, the X-Clusive 2388 and 2366 used an updated rotar, which increased capacity in high-yield, high-moisture crops that were particularly prevalant in northern Europe. The models were built from 2003 to 2008, and were replaced by the 88 series machines. *Case IH*

FRENCH RED COMBINES

By Jean Cointe

In 1951 the Ris-Orangis facility near Paris assembled the first pull-type combine—the F64. Production continued until 1953, when F64 production, as well as production of the F44, moved to the Croix factory. Ris-Orangis, meanwhile, became an important spare parts depot and a training center and later the home of the IH Payline division.

In 1955, 1,200 self-propelled combines and 1,700 pull-type combines were sold in France. Of the latter, Cima IH reached 50 percent market share despite issues with the F64.

The following year, 1956, saw the introduction of the F141 combine, an American-made machine imported from the East Moline plant and adapted to the French market with a Hispano-Suiza diesel engine and a different grain tank and sheet metal. Final assembly occurred at Croix.

The same year, Cima IH contracted with a French manufacturer, Merlin in Vierzon (which would become the future industrial site of J. I. Case France), to build a self-propelled combine, badged F8-83. This deal was aborted in 1958, leading to the bankruptcy of the supplier.

In 1957, a new self-propelled combine, the F8-61, came rolling off the assembly line at Croix. This machine was developed on the same engineering project as the German D8-61 and would be replaced by an upgraded version, the F8-63, in 1959.

With the advent of the European Common Market, the French subsidiary was chosen to head the development of a

Model F64

▼ The first harvester-thresher made in France by International Harvester featured engineering inspired by the American pull-type model 52. It was produced from 1951 to 1959 and available with a bagging platform or a 25-bushel grain tank as an option that allowed the user to replace the PTO drive with an auxiliary FC-123 engine. *Jean Cointe Collection*

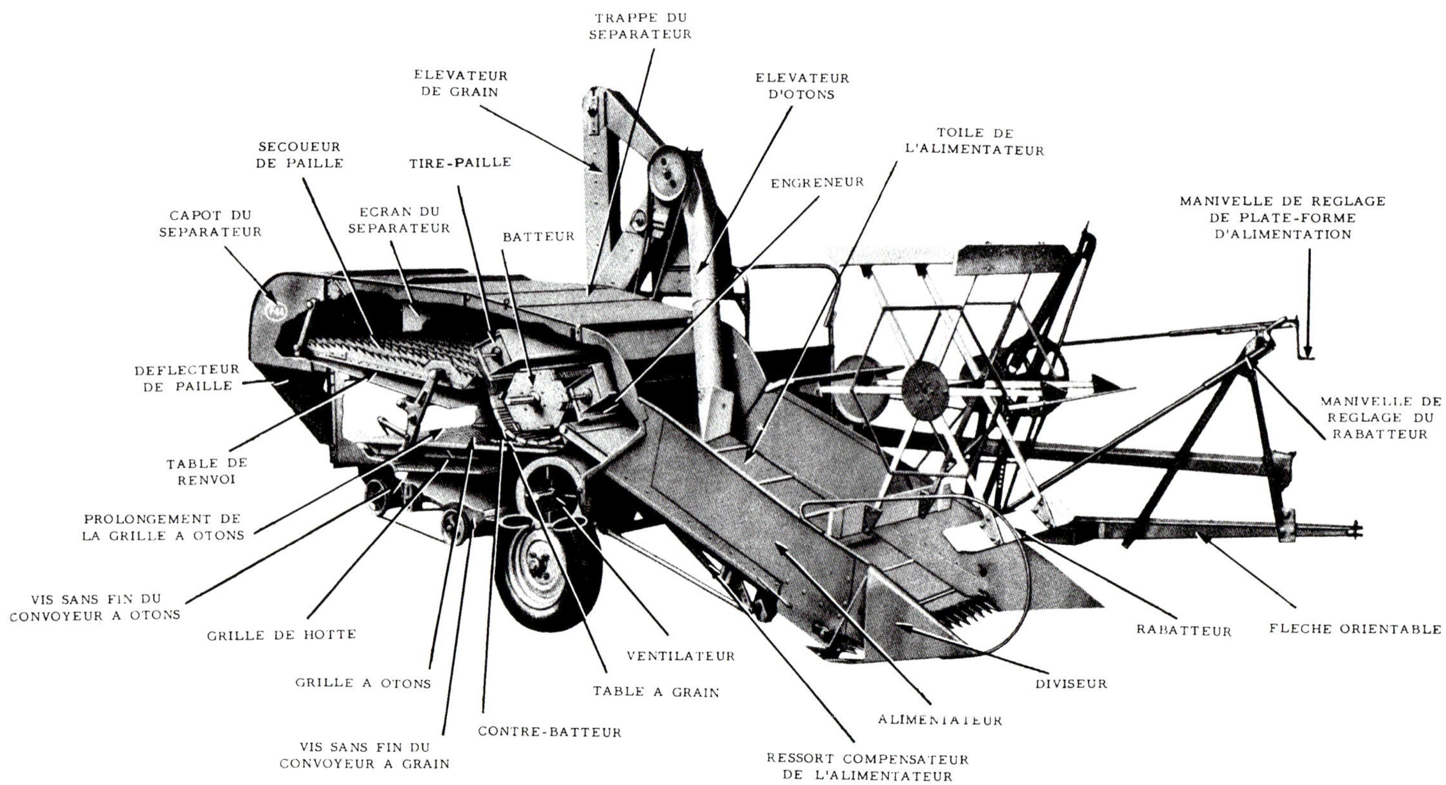

Model F44 Pull-Type

▲ This machine was unveiled in 1953 and produced at the Croix factory (Nord). An undersized machine, the F44 was equipped with a swivel hitch that didn't exceed 8 feet. The machine came standard with a bagging platform; a grain tank was optional. Production lasted through the 1959 model year. *Jean Cointe Collection* new range of combines. The first model, badged E8-41, was unveiled in 1963 with a modern design. At the same time, two imported American models, the 403 and 503 harvester-threshers, were assembled in the Croix facility with European specifications and rebadged the F8-413 and F8-513.

In 1969, in order to match the needs of the different crops found in Europe, new combines were added to the European line: the 8-51, 8-61, 8-71, and 8-91. These models were superseded in 1973 by the improved 221, 321, 431, and 531.

In 1977, the 541 foreshadowed a future range of new machines. The same year, the unveiling of the 953 combine, the first European IH machine equipped with a hydrostatic transmission, was followed by the downsized Models 943, 933, and 923. The American Axial-Flow combines were introduced in 1980.

In mid-1983, IH France purchased the Braud facility in Angers. Founded in 1870, Braud was the first French combine manufacturer. In 1984, a majority of the company was purchased by Fiatagri. Today, the firm is a subsidiary of CNH and the leading builder of self-propelled grape harvesters.

For more than two years, the modern Braud plant saw assembly of the European 1440, 1460, and 1420 combines with components delivered from East Moline. IH executives had announced in 1984 that the facility would be exclusively in charge of 1420 manufacturing worldwide, with the 1480 model also built there for export (except to North America).

Unfortunately, this Braud factory never reached the target of 1,500 machines per year. After the Case IH merger, the factories at Croix and Angers were closed in 1985 and 1986, thus sealing the fate of European red combines.

F8-83 Self-Propelled

Initially, this machine was outsourced from the French thresher company Merlin. After that firm's bankruptcy, it was probably assembled in Croix for a short period at the end of its 1957–1962 production run.

Jean Cointe Collection

F8-68 with FU-265 Tractor

The F8-68 was the last pull-type machine produced in France. Production was discontinued a year before the last F8-44 was built.

Jean Cointe Collection

F8-61 Self-Propelled

▲ Sharing identical engineering with the German D8-61, the F8-61 featured an FC-123 gasoline engine, a standard bagger, and no grain tank. Leading the way for the F8-63, the F8-61 was not a popular machine in France, but represented the first attempt by the Croix engineering team to develop a European IH harvester-thresher. The F8-61 lasted for just the 1957–1959 model years. *Jean Cointe Collection*

F8-44 Pull-Type

▲ The F8-44 was IH's last pull-type machine produced in France and featured upgrades including a hydraulic-actuated grain platform lift, a "free wheel" on the PTO shaft, a stone tub, and a new conveyor canvas. The French market by this time was strongly oriented toward high-capacity self-propelled combines, and production, which began for the 1960 model year, was discontinued with the 1964 models. *Jean Cointe Collection*

F8-63 Self-Propelled

▶ A grain tank, a bagger, and a combined grain tank with a bagger were available for the F8-63. Equipped with an IH D-132 or BD-144 gasoline engine, or an FD-123 diesel, it also had available grain platforms of 7- and 8-foot widths. *Jean Cointe Collection*

F141 Self-Propelled

▶ Imported from East Moline, the F141 combine could come with an optional Hispano-Suiza diesel engine in lieu of IH's Silver Diamond gasoline engine. Two grain-platform widths (9 and 11 feet) were also available for the French market, as was a bagging platform. The grain tank was standard. It was manufactured and imported for the 1957 and 1958 model years.

Jean Cointe Collection

F8-151 Self-Propelled

▲ Two engines were available for the F8-151: the Silver Diamond 240 gasoline model and the D-282 diesel. These machines were imported to fill a hole until suitable European machines were developed. *Jean Cointe Collection*

E8-41 Self-Propelled (1963-1965 Croix)

▲ The first "E" combine was an all-new machine specifically for the European market. It represented a new step for the Croix engineering team, which worked with the company's other research centers in its development. Manufactured for 1963–1965, the E8-41 was followed by a complete range. *Jean Cointe Collection*

F8-413 Self-Propelled

▲ Essentially a French 403 equipped with the D-282 engine and different sheet metal and equipment, the F8-413 was delivered partially assembled from East Moline and completed in the Croix facility from 1964 through 1967. Similarly, the F8-513 was a modified version of the 503 equipped with the D-301 engine, and it went through the same import and assembly process as the 403. Legend has it these models were rebadged to avoid infringement on Peugeot patents that also had a middle "0" in their model numbers. *Arnaud Caron*

8-41 Self-Propelled

▲ The 8-41 was basically the same machine as the former E8-41, but with the all-new IH D-206 engine (introduced in 1967). Built for the 1965–1969 model years, this combine allowed IH France to increase its market share with a modern and competitive machine. Part of the production was exported. *Arnaud Caron*

8-51 Self-Propelled Combine

▲ An update of the 8-41, the 8-51 featured a hydraulic-actuated cylinder dimmer, a nitrogen shock absorber for the grain platform (which was available in 7- and 9-foot widths), a D-206 engine, four-section straw racks, and a choice of two- or three-row corn heads. The first machine in a new complete range of IH France combines, it was built for the 1969–1972 model years. *Jean Cointe Collection*

8-61 Self-Propelled

▲ The 8-61 combine was equipped with a D-239 IH Neuss engine rated at 2,500 rpm. Features were identical to those of the 8-51 except for the introduction of a new quick-attach system for the grain platform, wider 9- and 11-foot platform widths, and increased 64-bushel grain-tank capacity. Built for the 1969–1972 model years, like the 8-51, it proved a popular model. *Jean Cointe Collection*

8-71 Self-Propelled

▲ Also produced for the 1969–1972 model years, the 8-71 combine sported the same features as the 8-61, along with some improvements. The grain-platform reel speed was hydraulically actuated from the operator seat, and a sound warning alerted the operator to a plugged air filter. Grain platforms were offered in 9- and 13-foot widths, and new for 1971 was IH's six-cylinder D-310 engines as an option. A panoramic cab made by Desmarais France was also available. *Jean Cointe Collection*

8-91 Self-Propelled

▲ The 8-91 combine was a heavier addition yet to a successful IH France lineup. Built for 1971 and 1972, it featured five-section straw racks, an 85-bushel grain tank, four-row corn headers, 11- and 15-foot platform widths, and a D-310 engine for 1971 and a D-358 for 1972. *Wisconsin Historical Society #117293*

221 Self-Propelled

▲ The 221 combine was identical to the 8-51 but rebadged to match the 321, 431, and 531. Produced only in 1972 and 1973 as this market segment decreased in France, it had a threshing cylinder that was hydraulically controlled from the operator seat. A panoramic cab was also available. *Jean Cointe Collection*

431 Self-Propelled

▲ The 431 succeeded the 8-71 and featured a larger 1-foot-diameter threshing cylinder. In 1973, the combine's second model year, IH France offered two engines: a four-cylinder IH D-239 and a six-cylinder D-310. In 1974 only the six-cylinder D-358 was offered. This is a 1973 model. The combine was produced through the 1980 model year. *Jean Cointe Collection*

321 Self-Propelled

▲ The 321 succeeded the 8-61 and enjoyed a long period of production (1972–1981). This machine was popular with customers and included options like a chain-actuated threshing cylinder, a grain tank cover, a quick-attach system for the grain platform, and a panoramic cab. A kit with working lights was also available. *Jean Cointe Collection*

531 Self-Propelled

▲ The 531 succeeded the 8-91 and featured five-section straw racks, 11- and 15-foot-wide grain platforms, four- or five-row corn heads, and a D-310 engine. The 531 remained IH France's "admiral vessel" despite the fact it didn't really compete with competitors' more advanced models. Nonetheless, it lasted from the 1972 through the 1978 model years. *Jean Cointe Collection*

541 Self-Propelled

▲ The 541 combine replaced the 531 and introduced some new features. For example, it was the first IH France combine equipped with a rotary-screen air filter, and its unloading auger was the first with a hydraulic swing control. It was a competitive model with a short production life (1978–1980).

Jean Cointe Collection

923 Self-Propelled

▲ The 923 was unveiled at the fall of 1980 as the small model of a new range headed by the 953. The 923 was unique, retaining a straw rack with four sections. There was also an optional dust vacuum for the conveyor. Dry-type brakes and a fan that's speed was actuated from the operator's seat were the main differences from the prior models. It was produced for 1980–1984. *Jean Cointe Collection*

933 Self-Propelled

▶ Like the 943 model, the 933 was low-priced but had more spartan features. The unloading auger was not equipped with an hydraulic swing control, and it had dry-type brakes in lieu of the wet discs featured on higher models. Built for 1980–1984, it had a D-358 engine and grain-platform widths of 12 and 16 feet.

Jean Cointe Collection

943 Self-Propelled (1980–1983 Croix)

▲ The 943 was the best machine in the last IH France series—an upgrade of the 953 with five-section straw racks that were shorter than those of the 953. The 943 had a large control platform and all the standard features of the 953 except a hydrostatic transmission. *Jean Cointe Collection*

Michel Hendricks' 1982 943 Combine

▲ Michel Hendricks owns and operates a farm in the east of France specializing in livestock. A faithful Case IH customer and the former owner of a 431 combine, Hendricks bought the 1982 943 shown here despite the fact it was produced 30 years ago. During the summer of 2014, Hendricks harvested more than 125 acres without any difficulties. *Jean Cointe Collection*

1978 953 Self-Propelled

◀ Unveiled on June 24, 1977, alongside the new 955, 1055, and 1246 tractors, the 953 was the first IH combine that allowed IH's European dealers to compete successfully with other high-capacity machines on the market. Equipped with wet disc brakes; a hydrostatic transmission (optional in 1980); grain platform widths of 11 and 15 feet; and corn heads in four, five, or six rows, it was produced from 1978 to 1983. *Jean Cointe Collection*

1982 International 1440

▲ An American-built 1440 Axial-Flow combine is shown near Aix, in the south of France in 1982. *Jean Cointe Collection*

Case IH 1480

▲ An American-built Axial-Flow combine is seen in Brussels, Belguim, on June 21, 1985, during a celebration of the alliance of J. I. Case and International Harvester.

Jean Cointe Collection

European Axial-Flows (1983-1985 Angers)

▲ International 1420, 1440, and 1460 Axial-Flow combines were assembled from 1983 to 1985 in a modern facility near Angers, France, using components made in East Moline and at a few European facilities (especially for cab components).

Jean Cointe Collection

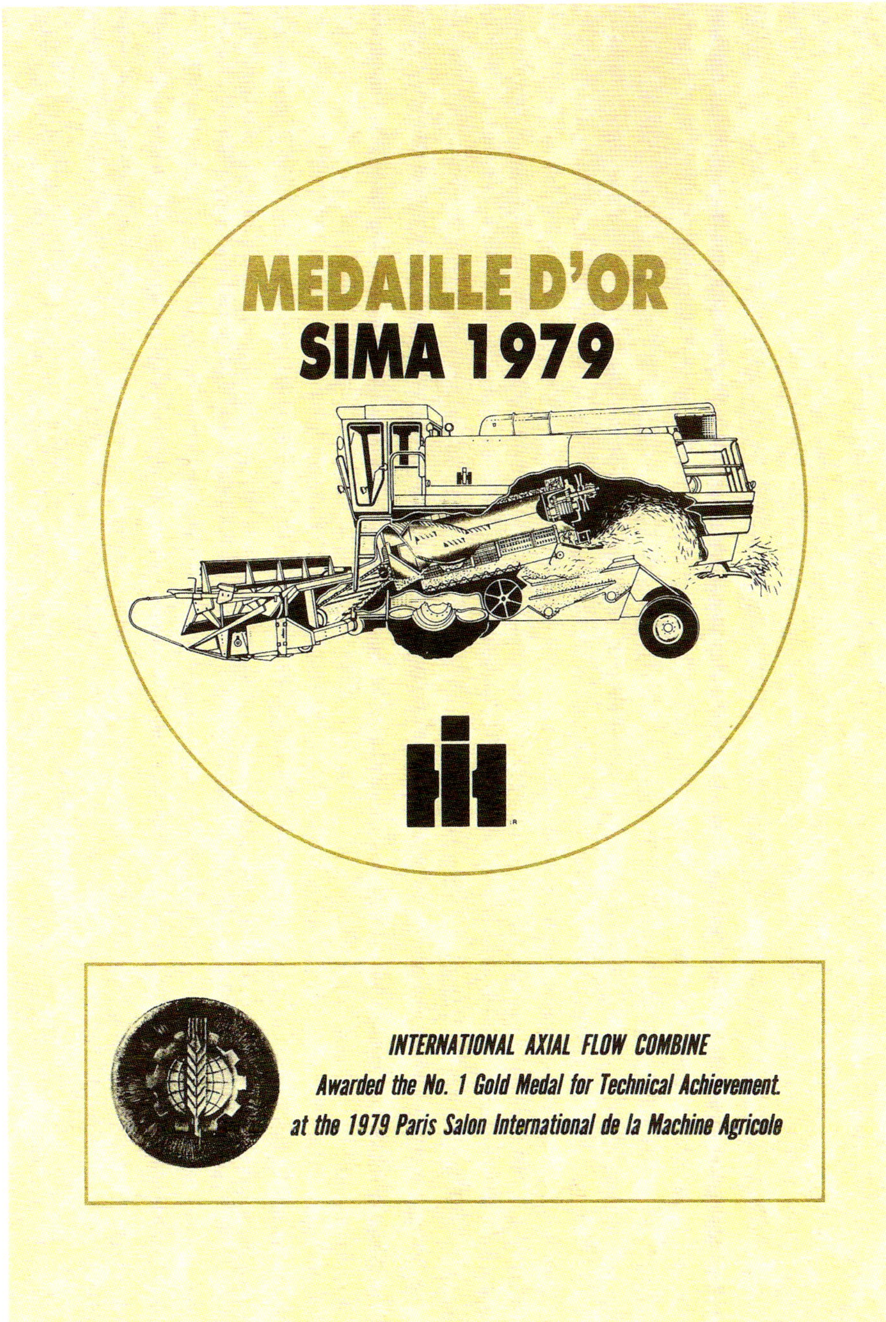

1979 SIMA Award

▲ The Paris International Agricultural Machinery Salon was the largest annual machinery display in the world in 1979, and the Axial-Flow combine was given the gold medal for its innovative technology and efficient field performance. *Case IH*

Engine and Chassis Assembly

▲ An Axial-Flow combine's engine and chassis are assembled at the plant near Angers, France. *Jean Cointe Collection*

Grain Tank Assembly

▶ An Axial-Flow combine's grain tank comes together at the Angers plant.

Jean Cointe Collection

1,000th Axial-Flow

▲ The 1,000th Axial-Flow assembled at the plant near Angers, France. *Jean Cointe Collection*

European Axial-Flow Combines

▶ The Axial-Flow combines built at Angers had a different access ladder. Most were sold with Model 825 European grain platforms.

Jean Cointe Collection

McCormick Combines in Sweden

▲ These images show a McCormick 6-foot pull-type and an 8-foot self-propelled combine built by Dronningborg and sold in Sweden. The pull-type is being pulled by a McCormick 624. *Jean Cointe Collection*

GERMAN RED COMBINES

By Matthias Buschmann and Johann Dittmer

First Harvester at Neuss Works

▲ International Harvester built harvesting equipment at Neuss Works, its plant in Germany, very early in the company's history. This is the first Harvester built there, on May 12, 1922. *Wisconsin Historical Society #17399*

Model D64

◀ These combines were available under the designation D44 and D64 with an auxiliary engine from either Volkswagen or an IHC diesel tractor, depending on customer preference. Also available was an accessory (built-on) hay baler. *Matthias Buschmann Collection*

Model D44

▲ While the French IHC sister company produced two tractor-drawn combines, the factory in Neuss procured both models from France and reconfigured them for conditions in Germany. *Matthias Buschmann Collection*

Models D8-61 and D8-62

▲ The self-propelled D61 arrived on the market in 1957, powered by the well-known VW Bug carbureted motor. The customer could pay for an upgrade to the IH diesel motor, which was built into the tractor. A built-on hay baler was also available as an option. Further developed technical improvements were introduced with the D8-61 and D8-62 beginning in 1959. *Matthias Buschmann Collection*

IHC President F. W. Jenks and the First Heidelberg Combine

▶ Due to the limitations of the IH factory in Neuss, the company acquired a manufacturing plant in Heidelberg. After renovations, combines built at this factory from 1959 to 1964 were supplied as "Made in Heidelberg." Though production of this line would later move to the plant in Croix, France, Heidelberg continued to serve as a construction equipment factory and produced the first Payloader line.

Matthias Buschmann Collection

German Axial-Flow Combines

▲ In 1980, Axial-Flow combines became available in Germany from East Moline and later from Angers, France. *Wisconsin Historical Society #117304 and 117306*

Dania 7200

▶ For customers who didn't want a rotary combine, Danish company Dronningborg produced straw-walker combines branded Dania and sold by Case IH from 1987 until the beginning of the 1990s. Marketed by Case IH and sold under the Dania designation, these combine lines offered power output from 75 to 220 hp. For cutting widths of 9 to 20 feet, the residual-grain separation system worked with four, five, or six shakers depending on the size of the combine. *Matthias Buschmann Collection*

The Neustadt-Built Case IH Line

▶ The 521 combine in Case IH paint. At the end of 1997, Case took over the distribution and production rights for the MDW combine factory in Singwitz, Germany, and the Fortschritt factory in Neustadt, Germany. This allowed Case IH distribution to make the straw-walker combines available for sale alongside the big baler, self-propelled forage harvesters, and Axial-Flow combines. Until then, the harvesters and balers had been sold under the names Mengele and Fortschritt, and the combines under the name MDW. This image shows the lineup in April 1998.

Matthias Buschmann Collection

MD Case IH 525

▲ *Case IH*

MD Case IH 527

▲ *Case IH*

Arcus 2500

◀ The 2500 was an innovative but flawed machine built by MDW. Originally painted blue, a handful were produced in Case IH red. With a 28-foot cutting width, an engine that produced 425 hp, and a 423-cubic-foot grain tank, the Arcus 2500 was a sensational machine in 1997. The Arcus 2500 combine was one of Europe's most powerful and innovative machines. The first Arcus combines were able to reach speeds of up to 25 miles per hour on the street. *Case IH*

Case IH CT5050

◀ The merger between Case and New Holland led to product standardizations. The harvesters, big balers, and combines under CNH were based on New Holland machines. In 2001 the factory in Neustadt assembled the NH TX construction line in red paint with the type designations CT 5050, CT 5060, CT 5070, and CT 5080. The closing of the Neustadt factory in 2004 brought the end to production of the CT line combines as well as sales of walker combines from the Case IH brand. *Matthias Buschmann Collection*

AUSTRALIAN RED COMBINES

By Sarah Tomac

The first McCormick reaper in Australia arrived in 1852—21 years after Cyrus Hall McCormick demonstrated his machine to a skeptical gathering at his father's farm near Steele's Tavern, Virginia.

Two separate organizations, one selling McCormick and one Deering, formed in Australia in 1884. The American merger of these two companies, along with a handful of other major manufacturers of agricultural equipment, occurred in 1902, and in 1903 International Harvester Company of America (IHCA) registered for trading in Australia. The following year, a joint venture between McCormick and Deering saw the resulting company selling not only their own farm implements, engines, and vehicles, but also acting as Australia's sole agents for Buffalo Pitts farm engines and thrashers, Chattanooga reversible disc plows, Cockshutt plows, Deere & Company steel plows, Oliver plows, Sanders disc plows, and Zealandia milking machines.

IH Stripper-Harvester

▼ A farmer uses a horse-drawn International stripper-harvester in an Australian field on June 21, 1920. *Wisconsin Historical Society #44771*

The company headquarters in Melbourne were established in 1904 for agricultural trade. All International Harvester goods were sold from this building, as were its Osbourne Division, Deere, and Buffalo Pitts agency lines.

IHCA occupied many other buildings in and around Melbourne as its needs expanded. The company assembled most of its components in Spotswood, a small suburb just outside Melbourne, on Hall Street. The company sent imported implements and tractors all over Australia from this location until Geelong Works opened in 1939.

COMPETITION

Agriculture equipment has been manufactured in Australia since the very beginning of the Agricultural Revolution. In 1843, John Ridley of Hindmarsh (near Adelaide), South Australia, developed a stripper, and in 1885 H. V. McKay of Sunshine, Victoria, produced a harvester stripper. IH competitors had Stump Jump disc cultivators and plows, as well as combines (seed and fertilizer) and header harvesters, in production by 1917.

IHCA's success importing machinery lasted until the worldwide economic depression of the 1930s, when the Australian government imposed heavy import duties on farm machinery. As this was IHCA's main trade, sales were affected greatly, resulting in renewed efforts to establish its own factory.

Construction was soon underway after the first sod was turned on July 19, 1938. Just months after construction started on Geelong Works, the first iron was poured at the foundry in December 1938. The plant officially opened May 22, 1939. Geelong would boast the greatest diversity of products built by any of IH's 44 manufacturing locations around the world. The site, located between the old Princes Highway and the new highway, is bounded on the south by the Melbourne-Geelong railway.

Melbourne's Harvester House

◀ A receptionist sits at her desk near a Farmall tractor in the entryway of "Harvester House" in Melbourne, Australia, in 1947. *Wisconsin Historical Society #45808*

WAR YEARS

Early in 1940, Geelong received its first military contract. The Commonwealth government had exercised its right to utilize the services of the plant and its workforce. Geelong Works became a large contributor to the war effort, manufacturing guns, tanks, shells, aircraft, and other war equipment. Geelong Works still continued to produce certain lines of farm equipment to help feed the thousands of soldiers overseas. Engineers at the works were able to develop equipment for large-scale vegetable production during the war, and as a result, many new lines of vegetable equipment were available in Australia.

GL200 and Farmall M

◀ In 1851, gold was discovered in the fields of Ballarat, Victoria. Roughly 100 years later, a golden harvest was taken instead.

S. Tomac Collection

At war's end, Geelong Works reverted to agricultural manufacture, determined more than ever to provide Australian farmers with everything they needed. IHCA realized that Australia had huge potential and was poised for a gigantic industrial revolution.

In the late 1930s, the company entered an agreement with Gaston Brothers, an Australian company that built header harvesters. The Lite-Draft Harvester was an Australian-made machine suited to the country's unique environment and conditions. The Gaston-built header harvester quickly became the basis for the GL 200 harvester.

In 1947, just a few short years after the expansion of the Geelong Works, the first IH-built header harvester rolled off the assembly line. The GL 200, a PTO machine, enjoyed huge success and was the mainstay for International Harvester Australia (IHA) from 1947 to 1957. The header had a grain tank capacity of 30 bushels and a 10-foot cutting width. An optional bagging platform was available. The GL 200 increased the Australian farmer's productivity immensely.

Potential for the expansion of the manufacturing line was promising in the 1950s. The purchase of 104 acres of land in Werribee, Victoria, for a new farm equipment facility was

International A8-1

▶ The A8-1 PTO machine was available with a 12- or 14-foot-wide attached cutting bar from 1959 to 1961 and featured 10 grease points. The list price in 1959 was £1,575, and 3,134 units were built in the three-year period before it was replaced with the A8-4. This one is shown in January 1960.

S. Tomac Collection

exciting news. The proposed Farm Equipment Works was to have machining, sheet metal, welding, fabrication, painting, and assembly facilities, as well as a warehousing area and loading docks. The Werribee Works were never built.

PRODUCTS

The Australian-built combine was slightly different in its conventional design, and although the drums, concaves, and straw walkers were based on an American 03 Series machine, significant changes were made to the components to allow them to better work in Australia.

During a 35-year period, IHA designed, built, and tested 11 different models. Field testing usually occurred in many different small rural areas, including Deniliquin, Ouyin, Horsham, Jung, Hopetown, St. Arnaud, and Wudinna. Testing of each machine—such as the A8-3—required six handbuilt machines to be shipped to the different states. Conditions vary drastically from the far north to the south of the country, and the header needed to perform flawlessly in every area.

An extensive trial of the A8-1 header on the wheat trail by the IH engineers proved the machine was ready. A team of IH employees headed by chief engineer George Minns took a convoy of trucks to the Moree district in northern New South Wales in 1958 with plans to test the machine with other products under a variety of conditions.

Over the course of 19 weeks the team harvested crops that were abundant and sparse, ranging from 4.5 bushels per acre to 66 bushels. Total acres harvested in this trial were 3,335; total bushels were 23,345. Highlights of the testing were the ability to field suggestions from the farmers and harvesting 107 acres in 14 hours with one machine.

Australia preferred the more traditional PTO machines in many farming areas. This may have been mainly because of cost and capacity. In 1967 a new A8-4 PTO header harvester, a basic machine without grain tank and a 14-foot front, listed for $4,458. The 135-bushel tank and auger unloader option was an extra $850. The basic self-propelled model, the 8-5 header harvester with an 85-hp engine, 72-bushel tank, and 14-foot open front, listed for $9,258. The American-built 503 harvester was also available in Australia in 1967. This 106-hp, 20-foot-cut, 110-bushel tank machine listed for $13,516.

The Australian way of harvesting was turned upside down with the 1979 introduction of the Axial-Flow. Until then, improvements to the machines focused on the concave, finger rake, and straw walkers, as well as cutting width, holding capacity,

International A8-2 Field Day

▲ Neil Murray, product engineer, spoke about the many new features of the A8-2 to potential owners and dealers at a field-day trial in Byford, Western Australia, in 1959.

S. Tomac Collection

International A8-2

◀ A caravan of A8-2 headers is transported undercover during the final stages of testing in 1958. The trial covered 6,000 road miles and lasted 19 weeks.

S. Tomac Collection

International A8-3

▶ The Australian-built self-propelled A8-3 was lever-driven, much like the American 91 model. The concept was similar, with chain drive and levers instead of a steering wheel. The next self-propelled model reverted to the traditional steering wheel.

S. Tomac Collection

International A8-4

▶ The A8-4 was a PTO-driven machine with an extra-capacity grain bin and unloading auger attached. From 1962 to 1973, 4,920 units were built in a comb width of 12 or 14 feet.

S. Tomac Collection

and, on the self-propelled models, the motor and the fuel tank. Demand for the Australian-built machines was declining; in 1979, there were 1,055 new Australian-built units. By 1981, only 479 were made. This was due in part to the decline of IH as it was known, and in part to the success of the Axial-Flow.

The introduction of the 1400 Series Axial-Flow to Australia was phenomenal to the farmer. Removing 16 parts of a conventional machine allowed for a more compact design and increased harvest capacity, lessened grain loss and damage, and kept maintenance simple. The operator was not forgotten either. The cabin offered the best in comfort and ease of usage. With a decibel level of 79, this was a very quiet machine compared to the competitors, which were still exceeding 110 dBs.

Many hours of training for the dealers and IH executives occurred before customers even knew the Axial-Flow was available. In fact, it had been tested and improved upon in Australia since 1966. It was originally designed for the better harvesting of corn and soybeans (more than 60 percent of U.S. crop sales are represented by corn and soybeans), but Australia is predominantly a small-grain country. A major factor in the Axial-Flow's success in Australia was the ability to easily clean the machine and move on to a different crop.

Harvest conditions in Australia are some of, if not *the*, harshest in the world. The Axial-Flow gave the farmer the ability to harvest faster and keep the clean grain with minimal loss. A majority of the Australian market is in the growth of high-protein white wheat. Sensitive to the moisture of rains (or even a heavy dew), the wheat needs to be harvested faster and more cleanly than wheat elsewhere in the world. Growing in a naturally dry country, the plant itself has a much shorter stalk than varieties in other countries. As a result of the close ground work, a lot of fine dust and dirt is brought into the machine, causing a much higher wear rate on the machine's parts. It follows that Australians work a machine much harder than anywhere else, harvesting at a higher ground speed and testing the durability and longevity of machinery. Testing the Axial-Flow in the harshest conditions known resulted in a better product worldwide. Many of the adaptions made in Australia were carried over to the final production.

One demonstration in the Moree area of New South Wales caused some disbelief for one farmer. Conventional machines damage a lot of final product, and he was having a hard time comprehending the concept of undamaged grain. The IH team

International 8-5

◀ Photographed in March 1965, this preproduction 8-5 with experimental tracks shows the height capabilities of the front, allowing for clearance through narrow gates and better harvest capabilities. Tracks instead of traditional tires allowed for better floatation and less ground compaction in rice fields. *S. Tomac Collection*

convinced him to take two loads of grain to the silos; the silo guys would report back how much crackage was in each load. The report came back, "It is impossible not to crack grain; therefore, we've made half of 1 percent crackage." In reality, there was zero crackage.

Geoff Rendell of Case IH Australia was one-half of the two-man team that decided to bring the Axial-Flow into the country. The two men were flown all over the United States in 1978, beginning at East Moline for service training and continuing on to wherever the machine was working to see it in action. If they didn't like what they saw in the Axial-Flow, they planned to go to France and inspect the combine that was being built there. The team never made it to France.

International 8-5

◀ The 8-5 handled high-yield wheat crops with ease, as was evident in the harvesting of Pinnacle wheat on Mr. T. Cousens' property about 20 miles north of Geelong, Victoria, in December 1966.

S. Tomac Collection

International 8-5 1967

▲ Fronts were built in three widths–12, 14, and 16 feet–and came with a choice of comb or open front. The header featured a larger cleaning fan and riddle box, a 72-bushel grain tank, and an 85-hp six-cylinder gasoline engine. The two-row corn front was also an option. *S. Tomac Collection*

In 1979, the total number of Axial-Flow machines in Australia was 13. Anyone interested in ordering the combine could call one of those 13 farmers and get an honest opinion. It was the start of a decades-long reign as the number-one combine sold in Australia.

Under IH policies, the Australian company had no restrictions on marketing its products in any part of the world, and Australian-made products were used in more than 90 countries. The same IH policies allowed facilities manufacturing IH products in the United States, Europe, and any of the 44 worldwide facilities to market and sell their products in Australia.

With Geelong Works closed, the plan was to import machinery and headers as needed, as well as sell the extra stockpiled products. The smaller population of Australia in the 1980s allowed this transition to work effectively. The market for headers in Australia required a little more horsepower and a larger concave. At the time, the 1400 Series in France was able to meet this requirement with the 1440 and the bestselling 1460 that used the DT-466 engine (the 1420 was still imported from the United States). Because this 1400 Series, with its increased horsepower and larger concave, was basically the same as the 1600 Series that was introduced in the United States, the 1600 Series machines were not introduced in Australia at the same time they debuted in the United States.

With the merger of Case and IH in Australia in 1986 (two years after the American merger), the company's combine market share dropped to 11 percent. During the transition, the whole dealer network was rearranged—some dealers were

International 8-5 Cab

◀ The first time a cabin was offered on an Australian-built header it was as an option on the 8-5. This one is pictured in April 1968. *S. Tomac Collection*

Cane Harvester

▲ A marketing agreement signed November 1968 with mechanized sugar-growing equipment manager Toft Brothers of Bundaberg, Queensland, gave IH better sales potential in cane-growing areas. The line of Toft sugarcane harvesters and loaders were marketed with the help of International Harvester Export Company headquartered in the United States. The highest sales of this equipment were to South America, Asia, and the United States. On September 1, 1970, Toft and IH announced that Toft would resume the distribution of its own line of cane-harvesting and handling equipment.

S. Tomac Collection

International 403

▲ This 403 was photographed in 1965 for advertising in Australia.

S. Tomac Collection

closed down and others were awarded the franchises. Once the network was stabilized, the company slowly began to regain market share.

It was soon apparent that training was necessary to increase market share. Many of the dealerships were unfamiliar with the products. Training focused first on service and then on the product. With the knowledge of the service available, dealers were better able to sell the product.

Phil Moore became Australia's managing director in 1989 and brought with him an ambitious strategy—he wanted the company to be number-one in every product. When his 10-year plan was presented to the management team, he told them if they weren't willing to achieve this goal they were to leave then and there. No one left. They all knew it was going to be an exciting ride.

The introduction of the Magnum allowed Case IH Australia to go head to head with Deere after everyone had full power-shift training. Other big areas were hay, cotton, and tillage equipment. With the basics set up, the ride to the top began.

The team knew how good the Axial-Flow was. Moore wanted a 3 percent sales increase year after year. By the end of 1989, the company held a 22 percent market share. At the end of 1991, it held 45 percent, and in 1999 it still held about 44 percent. Number one most years in tractor sales was the Magnum. Case IH combines held the number-one spot for 10 years straight from 1990 to 2000.

Key factors to holding that position for such a long time were the stabilization of the dealer network and intense product and service training. Another key was in the rebuttal to a letter written by a competitive dealer about the supposed failure of the Magnum as a tractor. A major marketing campaign was

International 8-6

▲ Built from 1968 to 1976 with a total of 1,835 units produced, the PTO-driven 8-6 featured a 14- or 16-foot cutting width and replaced the popular A8-4. This one is shown in November 1967. *S. Tomac Collection*

International 710

◀ Taken in May 1978, this overhead-view advertising photo shows how the 710 PTO offered a wider cutting width than previous models with the introduction of an 18-foot cutting bar. There were 1,082 units built from 1977 to 1979. *S. Tomac Collection*

launched refuting all 52 items listed in the letter. A representative and a Magnum tractor were sent around the country to discuss the tractor's qualities. The move showed that Case IH was willing to meet challenges head on.

Through the last 24 years, there have been just five years that Case IH hasn't held the number-one spot in Australia, and the company has been able to recover from any slumps they experienced. Australia is a long way from anywhere, and importing equipment offers lots of opportunity for disaster to strike. Products arrive by ship, and tractors are packed in containers with major and minor components separate. They are assembled and tested when they arrive at the main warehouse. Combines, however, are too big to send in a container and are often driven into the understorage of the cargo boat. Exposed to the elements, machines can, and often do, arrive scratched, dented, rusted, and in need of some cosmetic assistance to make them ready for sale. Sending the machines back to the factory for repair work would not be cost-effective. Communication between the factory, dealers, and customers has minimized the damage caused on these long journeys to their final destination.

The 2100 Series became one of the best models available to the Australian farmer. By 1999 competitors held a small share (Deere with 23 percent and Massey, New Holland, and the rest sharing about 14 percent—Case held the rest). Despite the merger of Case and New Holland in the United States, the two companies remained completely separate in Australia. Some market share was lost in 2000, but by 2002 Case was number one again.

International 711

▶ The 711 replaced the flagship 8-5. Built from 1973 to 1978, the model was the first to offer a diesel engine in the Australian-built machines, as well as a range of comb sizes: 16, 18, and 20 feet. It was the most popular self-propelled header built at Geelong Works. *S. Tomac Collection*

The Old and the New

▲ A full-size 711 sits next to a scale model of a 725—the model that replaced it. This image was taken in February 1979. *S. Tomac Collection*

Experimental 725

▲ The crew testing an FX-17, the experimental 725, stops for a break and a photo opportunity in December 1977. The 725 was released in late 1979 and built until 1981. *S. Tomac Collection*

Competition between Case IH and Deere has always been fierce, and leadership has swapped nearly every year since 2004. The 8010 was introduced in 2004. That year Case sold 120 combines in comparison to Deere's 104 units of its 9760. Deere remained on top that year with the carryover of 9860 sales, which were only a few units higher than sales of the Case 2388. Sales are always close and monitored monthly, as seven machines could mean a 1 percent difference.

The salesman has to be well-trained, sell himself, and know the product well in order to keep the market leader position. The combination of dealer support, product support, and sales support makes the Case IH team a market leader year after year.

FX-17

▲ The FX-17's control center is seen in November 1977. From 1979 to 1981, IH built 1,310 units of Australia's 725 self-propelled header. *S. Tomac Collection*

725 in Melbourne

▲ Part of the marketing campaign to enlighten the urban population about the advances in the production of food included displaying the new 725 in Melbourne on Bourke Street in 1980. *S. Tomac Collection*

1979 International 725

▲ A side view of the scale model 725 in 1979. The addition of full sheet metal to cover the exposed moving parts and the introduction of a quieter cabin with standard air conditioning gave the new model a modern look. *S. Tomac Collection*

International 725 MKII

▶ The last 33 units that were built at Geelong Works were renamed the 725 MKII for the 1982 Australian harvest. With a marketing campaign titled "The Staying Power," IH Australia was not going to let financial disaster interrupt its promise of delivery to the customer. *S. Tomac Collection*

International 726

Final stages in the buildup of the 726 PTO header. Sold from 1980 to 1982, only 725 of these were built before Geelong Works closed. *S. Tomac Collection*

International 726

A 726 is towed by an 886 tractor in a 1980 field trial. The 726 was the last header produced at Geelong. *S. Tomac Collection*

International 915

◀ This 1975 photo shows a 915 bearing the name "Grain and Maize Special" on trial in Australia. The diesel turbo was unlike anything Australia had available at the time.

S. Tomac Collection

Geelong Works

▲ A view of the Geelong Works in late 1970s. *S. Tomac Collection*

Work Floor

▲ Finishing touches are put on key components in Geelong Works.

S. Tomac Collection

1460 in Northern Victoria

Introduced in Australia in 1979, the 1460 was easy to clean and adjust, allowing the farmer to harvest a wider variety of crops in a faster, cleaner, and more reliable fashion.

S. Tomac Collection

International 1460

This 1980 photo of a 1460 on the lawn of Geelong Works shows the open front head without reel, which was in common use in Australia. *S. Tomac Collection*

RED COMBINES IN CENTRAL ASIA

When Partice Loiseleur was tasked with convincing leaders from Turkmenistan that red combines provided more longevity and value than the much cheaper Russian-built equipment being used at that time, he turned to a farm family from Rochelle, Illinois. The family had farmed for three generations on the same piece of ground and used nothing but red equipment. They also maintained their machines and kept them running for decades.

The farm was on the route Loiseleur frequently used when taking foreign groups on tours. While en route between manufacturing facilities, Loiseleur would stop with his group at this farm. The questions he would ask became a familiar routine.

"The joke with the rep and with the farmer was to go into the oldest tractor in the field and ask the farmer, 'How old is this tractor?'

"'It is 17 years old.'

"How many hours do you have on the counter?

"'We must have 10,000 hours.'

"How many times have you opened the engine?

"'Never, why would I need to rebuild the engine after 10,000 hours? It's brand new!'"

The representatives from Turkmenistan reacted with shock.

"Obviously, they expected less and could not believe what they heard because for them, an engine couldn't have lasted for more than 1,500 hours," Loiseleur said. "That was the kind of method. We have to convince them about our machine reliability, our machine performance compared to what they have been using so far."

The bit proved to be key, as Case IH equipment cost seven times more than the Soviet equipment being used in much of Central Asia. The area was of great interest to Case IH in the early 1990s.

When the Soviet Union collapsed in 1991, the Turkish-speaking countries of Central Asia found themselves free for a time. The area is rich in oil, and the harsh climate and poor transportation infrastructure limits imports. The region relies heavily on internal agriculture to feed its people with grain and clothe them with cotton.

With the borders open and the existing equipment at the time being Soviet-built machines that lasted less than 1,500 hours and harvested poorly, opportunity existed for an equipment manufacturer to enter the market.

Loiseleur headed the sales effort to Turkmenistan, the first country Case IH approached. The task proved tremendously challenging. While the company had ties to Russia dating back to the 1910s, Case IH had no existing network, dealerships, or even sales history with Central Asia.

"There was no dealer, there was no service infrastructure. There was no, nothing in between," Loiseleur said. "The challenge was not only to provide the machines, but also to provide everything around the machine, meaning that we had

Central Asia

▼ The Caucasus is a region between the Black Sea and the Caspian Sea. The region features a number of Turkish-speaking countries, most of which were part of the Soviet Union until 1991. When the Soviet Union dissolved, many of these countries became independent. The region has rich natural resources, limited transportation access, and complicated politics. *USAID*

The Power of Longevity

▶ When dignitaries from Central Asia were brought to America to see red equipment in action, they were amazed that the machines could last for generations. This 1986 Case IH 1660 wears original paint and in 2015 was still hard at work on the farm of owner Jeff Jacobs. *Lee Klancher*

The Door to Hell

▶ The flaming cavern known as the Door to Hell is located near the village of Darvaza in Turkmenistan. The area is rich in natural gas. When a drilling rig collapsed into an underground cavern in 1971, the escaping gas was lit in order to stop the fumes from poisoning the air. Geologists hoped the fire would go out in a few days, but the cavern has been spouting flames for more than 40 years.

flickr.com/flydime

to supply service, we had to supply the parts, we had to supply even lubricant."

They also had to build relationships with the government, and convince it that Case IH could provide all the service necessary.

All this was done without any immediate support from the head office in Europe or the United States. Calls to France could take up to day to get an answer. Loiseleur carried a printer, paper, brochures, and everything he needed with him, and had to provide answers on the spot.

"We could only rely on ourselves," he said.

While Loiseleur was on his own day-to-day, he enjoyed strong support for the program from Case Corporation leader Jean-Pierre Rosso. Rosso was pushing the company to explore and grow, and took a personal interest in seeing the Central Asia initiative succeed.

Deals had to be signed off on personally by the president of the country. Large banquets were part of these dealings, with fantastic buffets and heavy doses of vodka.

"Then obviously you had to drink vodka because it is in the tradition, when you have a big banquet like this, you have to do toasts," Loiseleur said. "Toast about the friendship, toast about the man-woman relationship, toast about the Case IH project, toast about the next wheat crop and wheat yield and so on. It could last for hours and hours.

"Then always, again, you would end in the sauna because it was part of the tradition."

In Uzbekistan, the first contract signed in Central Asia was for a handful of combines and cotton pickers. The terms were agreed to late in the afternoon. While the documents were prepared, Loiseleur and the officials went for dinner. Which led to vodka, toasting, and—as is traditional—a sauna.

"Late that night, another secretary came back with the document in order to get it signed," Loiseleur said. "I signed the first contract in Uzbekistan just wearing a toga around my waist and so was the minister of ag."

Those early sales—along with strong performance of the machines—led to large orders of 300 to 500 combines at a time, all over Central Asia. One of the first large orders was 300 combines to Turkmenistan. The first 100 shipped to the country—with massive effort as the infrastructure was new—but they arrived without issue.

The second shipment of 200 was another matter. At about the same time the shipment arrived in Central Asia, Loiseleur received a message informing him that the machines had a gearbox problem caused by a 50-cent part.

All the gearboxes would have to be rebuilt.

"For me, it was a huge catastrophe because for 18 months that I had been repeatedly saying that Case IH combines were the most reliable, that our machine could last 10 years and run 10,000 hours without any issue," Loiseleur said. "I had to go back to the minister of ag and tell him that we had to rebuild the gearbox even before commissioning the machine.

"We were scared to death about what could be their reaction, and the funny thing was that it was the best selling argument ever. It was the very first time a supplier came to them in order to mention a breakdown that may happen in

the future and that would be fixed even before it happens at our own cost. What we thought would have a negative impact on our business turned to be a very, very positive fact for us."

Language proved to be another barrier, and finding trustworthy interpreters was difficult. Loisleur's group found a Russian interpreter based in Chicago who would travel with them. He not only interpreted, but he also understood local customs and culture. He proved invaluable.

In Kazakhstan, Loiseleur recalled selling a test machine proved exciting. One of their local helpers found a buyer for the machine. When it came time to take payment of nearly $200,000, the team was in for a surprise.

"We were expecting to get paid by check or something," Loiseleur said. They went to the local farm, and a guy was there with a bag stuffed with nearly $200,000 in cash.

"At that time . . . people would kill somebody for $200," Loiseleur said. They were concerned that the whole thing was a setup, and someone would be waiting to ambush them on the road.

"We had to organize to hire police forces from the city in order to babysit him until he was far enough away from the city to be safe," Loiseleur said. "That was the kind of life we had."

Bound for Central Asia

▲ This Northern Suffolk train is loaded with 1660 combines on their way to Turkmenistan or Uzbekistan via the port of Baltimore.

Case IH

RED COMBINES IN LATIN AMERICA

The Axial-Flow combine first arrived in Latin America in 1978. The machine was imported by a Brazilian farmer. Edgardo Legar worked for Massey Ferguson in Latin America, and he saw that big red machine first-hand when it came to Brazil. As of 2015, Legar believed it was still working the fields.

Nearly two decades later, Legar would get a chance to bring in more Axial-Flow combines. In 1996, Legar spearheaded an effort to bring red machinery into Brazil in a more serious effort. In fact, he was the first employee in the agricultural office in Brazil.

For a time, he was the *only* Case IH agricultural employee in Brazil.

Everything needed to be done to establish the business. The only related presence in the country was the Case construction equipment line.

"Everything had to be done for agricultural equipment support, so I needed to hire people, hire training engineers, and train these people," Legar said. He also personally established new dealerships and did customer deliveries. The demands on his time in those early days were intense.

"Hundreds of customers at that time had my telephone number so I worked for seven days a week, 365 days a year," he said.

1923 Brochure

▶ International Harvester exported equipment into South America early, as evidenced by this South American advertising poster for the McCormick harvester-thresher. *Wisconsin Historical Society #10048*

Legar was joined by other employees fairly quickly. One of those was Jose Camargo, a longtime sales rep for Massey Ferguson in Brazil who came on as a regional sales manager for Case IH in October 1996. He was hired and told to focus on selling machines to the large Brazilian farmers.

Brazil has some of the largest farms in the world, with some operations working more than 100,000 acres. The strategy made sense.

A customer who found Case IH would change that direction, at least in the early days.

The first sale of Case IH machines in Brazil was to the Tiecher brothers, who farmed near Luziania. They had seen the vintage IH 1460 at work not far from their farm and had decided they wanted to eliminate the hand work of harvesting edible beans by using a 2166 combine. Three 2166s were delivered to the Tiecher brothers in December 1996, the week before Christmas.

Jim Minihan, a Case IH engineer, came down to help set up the machines for the customer. Christian Lancestremere was also there for the setup. He was a representative for Case IH in Argentina and knew how to deal with local conditions.

The machines represented monumental change for some Brazilian farmers. Many of them used an old system of planting and harvesting edible beans that required each plant to be pulled from the ground by hand, and then the separation was done mechanically by a machine pulled with a tractor.

Minihan and Lancestremere set up the 2166 so the Tiecher brothers could harvest their edible beans in one fell swoop. The setup was completely foreign to Minihan, and Camargo remembers the two having an epic battle about the proper way to set up the combines.

In the end, the machine was a vast improvement, but both the way the farmers harvested and the configuration of the machine would have to be altered before edible beans in South America could be properly harvested.

To end the hand harvesting, the farmers had to change from planting in ditches to planting on flat ground. Legar and his team had a book commissioned that explained the new techniques to the farmers.

An Axial-Flow combine was not just a step forward in technology, it was like traveling to the moon. "It was so new for everyone," Legar said. "People came from every place in Brazil to look at them."

Less than a year later, Legar remembered meeting a ship at the dock loaded with more than 200 Case IH machines. Roughly 30 of those were red combines—a great start.

He also recalled massive training required to create a team that could service and support the machines following into Brazil. One key solution was to create a team of trained service technicians that could travel by truck to do field setups and technical deliveries. The team was sent to America to learn, and a few experienced American field service support representatives also came to Brazil to help train.

The other side of the coin was modifying the machines to work properly in Brazil's harsh conditions. The crops tended to be wet, and the soils contained more sand and metal than was typically found in North America.

Growing Market

▲ Brazil is home to some of the world's largest farms, a few of which cover more than 100,000 acres. *Case IH*

Jose Veiga, from Brazil Product Engineering, worked closely with the engineering team based in East Moline, Illinois, to get the machines running properly in Latin America. Two of the big challenges were that the crops were harvested with higher moisture content, and the machines ran longer hours in warm conditions. Brazil had some very large farms and also has a very long growing season. In some areas, three harvests are possible each year.

The machines were modified with more horsepower, improved cooling systems, concaves that wouldn't clog with the wet material, and modifications to the rotor to make it more durable and better suited to local crops.

The soil in Brazil was heavily laden with sand and metals. Minihan famously took a magnet to the soil and had it come away covered in iron ores. The ores would quickly wear out the North American knife guards, threshing bars, and bushings.

High-wear items were developed for the Latin American machines, and several of those were later adopted for use worldwide due to their increased durability. Evolution of the combine is a never-ending process, and what's vital in one region often proves beneficial to another.

Local dealers like Antonio Franciosi, the owner of Maxum Maquinas, said working this evolution was critical. "We helped the factory develop [combines] by passing along any defects, what was happening, giving ideas," Franciosi said. "There is always something good that the factory can gain from this. Because I am both a farmer and a dealer, I can work together with the factory without leaking information or things like that."

First Latin American Axial-Flow Built

▶ This 2388 was the first Axial-Flow combine built in Latin America. *Case IH*

The beans would prove critical to the Axial-Flow combine's success in Latin America. Jose Vengrus works for Bom Futuro, one of the largest farms in the world, and worked extensively with Case IH to develop the bean harvesting system. With his help, special soybean harvesting packages were developed that included a redesigned rotor cage and other modifications that removed the dust from the sample.

Once the improvements were in place, dealership owner Gerson Luis Garbuio said the customers had to be convinced the modified rotary machines could handle the tough conditions in the area.

"The challenge was to convince the farmer that this machine harvested in dirty areas," Garbuio said. "At that time, the difficulty was to prove that the machine also harvested in dirty areas. Conventional machines worked very well in dirty areas."

Other dealers found that the strategy of going after high-profile large customers didn't work in their region. Carlos Alberto Da Rosa opened his Case IH dealership in November 1997, and his region had mostly small- and medium-sized farmers who worked hilly ground. He had to convince customers that the large machines would work well on their farms.

"The brand had a slogan that was more or less, 'for large customers and for unique customers,' for customers who were looking for technology," Da Rosa said. "Of course, our machine with a rotor was an advantage, and we sold it and still sell it as a new technology.... To those customers we had to show that the rotor was a better option that increased quality when harvesting grain.

"It provided a better quality of harvest with less mechanical damage and greater productivity per hour worked. Until today this is one of our focuses when we work with the client and show our products."

By 2001, Case IH had established a good base of dealerships and customers in Latin America, particularly in Brazil. The key to expanding further was building domestically.

Mirco Romangnoli came to Latin America in 1994 from Modena, Italy, and had a variety of key positions within CNH heritage group all over the world. He started with Case IH in 2001 as senior director of marketing, and by 2015 he was vice president of Case IH agriculture in Latin America.

In 2001, Romangnoli was focused on building machines in Brazil. This was key because products made domestically could be financed. Financing imports was difficult and expensive for Brazilian farmers. This limited the market to those who could pay with cash.

By building the combines in Brazil, the access to financing would increase. This also reduced headaches with shipping, and stock was closer to the dealerships.

Romangnoli also said that the directive from a sales perspective went back to the original ideas that brought Case IH to Latin America in the 1990s: focus on the large farmers.

Case IH developed a new model to do this. It would provide direct factory service and support to the largest farmers in Brazil, while the dealers would handle the service and support to the small- and mid-sized farms.

The marketing strategy was the same one that has been used to sell Axial-Flow combines around the world since the late 1970s: a heavy emphasis on field demonstrations. The market in Latin America was largely straw walker machines, at least early on, and the performance of the rotaries has transformed that.

"We took the large customer by storm," Romangnoli said. "We really spent many, many hours there out in the field testing the machine with the customer, and we could show the productivity of the machine was extremely higher than any other conventional combine, that the level of quality of breaks in the grains was basically close to zero compared to the conventional that they had, strong breakage in their grains, which is very important to the customer, but especially the seeds producer, and especially the level of losses of our machine was a lot smaller than any other conventional combine. So we were able to show these three basic concepts: productivity, losses, and grain damage where we basically were leaders in the market."

The combination of local manufacturing and focused customer service increased the Case IH market share by leaps and bounds, and Brazil became a strong market for Case IH.

In 2013, the decision was made to focus on growth in Argentina, and the effort was headed by Christian Lancestremere. He focused on growing the dealer network and construction of a new plant in Cordoba, Argentina.

The plant construction helped hold down prices of the combines due to Argentinian policies that tariff imports. Case IH also focused more on the average farmer in that market and offered the 2388 Special, which is a lower-priced version of the model designed for the farmer who is purchasing his first Axial-Flow combine.

Case IH 2799

◀ Combines for Brazil and Argentina are built in each respective country. *Case IH*

The new plant came online in 2014, and sales have grown steadily ever since. The result of the efforts has been a strong increase in marketshare in Argentina. Case IH has increased its presence from having 15 dealerships to 36 in 2015. Improving that dealer network continues to be a priority for the group.

Lancestremere said that Argentinian farmers are closely attuned to what's happening America. "Argentine farmers look very close to the technology that the farmers use in North America, very close. And this was and is one of the reasons that every single year, I take Argentine farmers and I go to North America in order to visit North American farmers, to visit our plants, and show them the latest advances," he said. Precision farming, for example, is of great interest to his customers.

According to Geraldo Criolani, general manager of one of the largest dealerships in Latin America, the flagship combine launch was a major initiative.

"Another important challenge turned up in 2006, when the Axial-Flow 8010 was launched in Argentina, since we had to adapt ourselves to this new technology," Criolani said. "This involved major training of our staff, especially in the area of services."

Today, Case IH has plants in Curitiba and Sorocaba in Brazil and Cordoba, Argentina. Argentina and Brazil make up more than 70 percent of the farm equipment market in South America, and the continent continues to be a key market for Case IH going forward.

Case IH 2566

▶ The conditions in Latin America are hard, and the machines require more horsepower and different setups to deal with the crops.

Case IH

Claiton Bervian started selling Case IH equipment at one dealership in 1999, and his company, Meta Dealership, had eight stores in the Rio Grande do Sul region.

In the early days, his primary concern was how to provide good service to new customers. "I remember a day when we had a serious problem with the gearbox of a harvester, and exactly on that day we were lucky that one of our employees was at the factory, on the assembly line, on the day this happened," Bervian said. "He got Case IH to approve sending an entire gearbox at the time to take to the customer's farm to change the part.

"From that moment on we realized that whoever had spare parts, whole sets, and was able to provide quick part replacement support, would win over the customer's confidence. Any customer, small, medium, or large, regardless of their size, deserves good after-sales customer support. The argument was that the machine is large, almost twice as big as others at that time, and if it stopped for a day, it could be considered as two machines being stopped. The longer the downtime, the greater the expense. Our difference is that we were able to quickly take care of a serious problem and gain customer confidence. He had bought his first harvester; today he has eight large ones on his farm.

"In fact, it was a great evolution. In the beginning we started with just one dealership, one point of sales. And over these 16 years we have seen the brand grow here in the region. Today, we have eight stores in Rio Grande do Sul. The evolution was very great."

Case IH 2688

▲ A variety of models are built for the local markets, each closely based on the machines built in North America. *Case IH*

Soracaba Plant

▲ The manufacturing site and distribution center for agricultural products including Case IH is located in Sorcaba, Brazil. The center employs 1,700 people. *Case IH*

RED COMBINES IN CHINA

In 2006, a man from eastern China imported a 2388 combine to his village. The machine worked well for him, so he bought a 6088 in 2009 and a 6130 in 2012.

The machines made an impression on his neighbors, and today, the village has 20 Axial-Flow combines.

This isn't the first time red combines worked Chinese soil. International Harvester exported machinery to China as early as 1909, and the company had a branch office that opened in 1921 and closed in 1937.

IH 1460 in China

▶ This model's hard work in China helped convince farmers there was value in Axial-Flow combines. *Case IH*

Case IH 4077

▶ This model is built in China, and the design is based on the North American–built 2366. *Case IH*

After World War II, International Harvester brought 20 Chinese engineering students to America to learn about agricultural engineering and machinery. These students came back to China and helped form the manufacturers that serve the market in country today.

In the 1980s, a state-run farm purchased a large amount of International Harvester machinery. Many of those machines are still at work today.

In 2009, Case IH began work to establish a market in China. Weihong Zhang has been directing the effort and explained that the market is huge, though the bulk of the sales are of small combines that produce 50 to 100 horsepower. More than 120,000 of these are sold each year, and that market is dominated by a Chinese manufacturer, Foton Lobol.

The opportunities for Case IH reside with the larger farms, particularly the corn farmers in the northern part of the country, a market in which the machines already have had success. The company is launching a new machine that is built in China and is based on the 2366. The models are the 4077 and 4088.

According to Sean Lennon, who is working with Zhang in China, one of the key elements of the new program is a fleet of 16 teams that use service vehicles to provide support in the field.

The dealer network also needs to expand, and competing in the Chinese market is not easy. "It's a supermarket, so they have many, many brands and more direct loyalty to the brands," Lennon said. "With this being a premium product, we're working to take different steps to offset this."

Other challenges include getting to the key decision-makers on deals, strict government regulation, and convincing farmers to take a chance and depart from tradition.

Chinese farmers have typically used different row spacing than the rest of the world, for example. Getting them to change has been tough.

"Then, from a row spacing point of view, they generally operate on 70cm, whereas U.S. is more 76," Lennon said. "Also, they plant on ridges rather than on flat ground, and this is when, from an agronomic point of view, it is very difficult to understand why they do this. This is just a practice that has always been done.

"If they change and it fails, then everybody will look at them for making a change and it's failed. They are quite conservative in their approach to especially agronomic changes."

According to Zhang, the low breakage and loss of the Axial-Flow combine gives them a good edge to convince the Chinese farmer to make a change.

"With our machine," Zhang said, "they have a very low loss. Normally the loss for a local combine is two to three percent. With the Case IH combine, it's below one percent. They can save 400 kilograms per year."

Case IH 4088

▼ The corn heads available on the new line of Chinese-built machines will be five- and six-row heads modified for use in Chinese fields. *Case IH*

AUSTOFT
MAGNUM

Chapter Ten

HARVESTING EQUIPMENT

By Lee Klancher

"Our deep respect for the land and its harvest is the legacy of generations of farmers who put food on our tables, preserved our landscape, and inspired us with a powerful work ethic."
—James H. Douglas Jr.

While combines are the star of the International Harvester and Case IH lines, a wide variety of other machines were built to harvest corn, rice, and a host of other crops. Everything from coffee to dandelions needs to be harvested in some fashion, and companies around the world built machines to do that.

This section focuses on the best-known machines and shows just a few examples of each one. The later machinery gets a bit more space. Each machine has undergone a century of development and played a role in transforming the farm into a mechanized factory capable of feeding the world.

Harvesting the World

◀ Today's farmers use a wide variety of equipment and techniques to feed the world. This Case IH 7700 Sugar Cane Harvester is at work in Latin America. *Case IH*

CORN HARVESTERS

As late as the 1930s, corn harvesting was done by hand. Farmers would spend hours husking ears, and speed in the task was a prized skill. An average farmer could husk about 300 ears an hour.

The skill was showcased in the National Cornhusking Championship, the first of which was held in Polk County, Iowa, and was attended by 800 souls who braved freezing weather to watch state champions from around the country do battle in the 80-minute contest.

The champion that year, Fred Stankek, was able to husk 2,000 ears per hour, an astounding feat. The 1925 contest was held in Illinois and drew a much larger crowd. By 1930, crowds of 30,000 people or more showed up to crown the corn king. By 1936, about 160,000 people came to the event, which was dubbed one of the fastest-growing spectator sports in America.

The competition was sidelined during World War II and never came back. Perhaps part of that was due to the fact that corn husking was no longer necessary, as first custom-built corn pickers were able to supplant the hand work, and these were later replaced by the corn head on the combine.

The Importance of Corn

▼ Corn was one of the most important crops harvested.
Wisconsin Historical Society

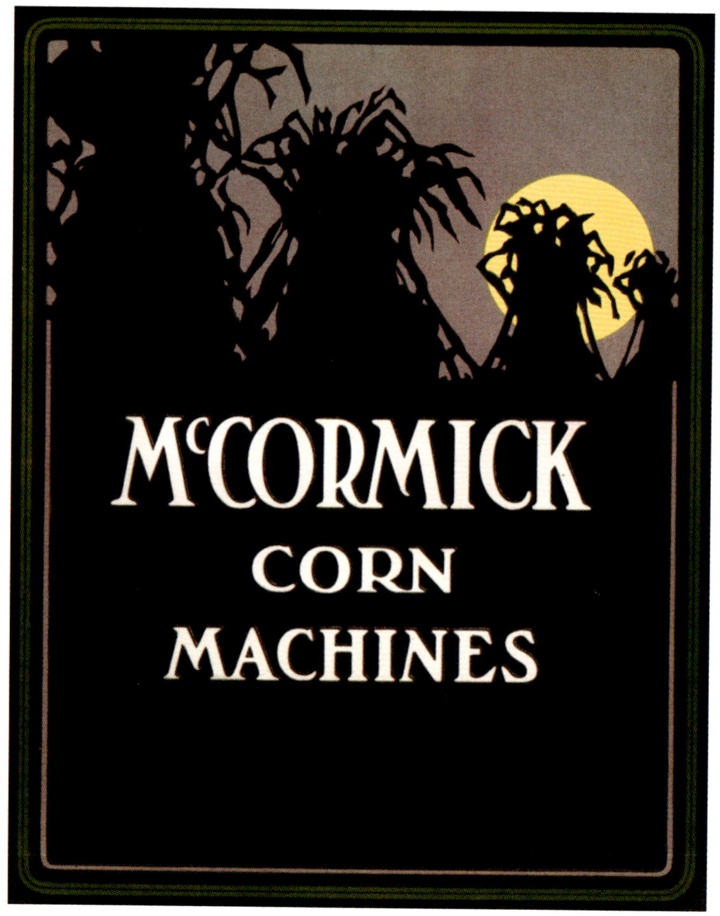

Dawn of Corn

▲ Cover of a 1916 advertising catalog for International Harvester's McCormick line of corn machines showing shocks of corn silhouetted against a moonlit sky. *Wisconsin Historical Society #23430*

Corn Binder Demonstration

▲ Demonstration for the 1915 Panama-Pacific Exposition in California showing how the labor swing bundle elevator of the Deering corn binder operates.

Wisconsin Historical Society #44696

1938 Farmall F-20 with 2-ME Corn Picker

▲ Pickers were offered by International Harvester that fit nicely on the company's F-20 tractor. Owned by Adam and Dan Robinson. *Lee Klancher*

1948 Farmall C with Model 24 Picker

◀ This 1948 Farmall C is mounted with a one-row Model 24 corn picker. Owner Brian Hale reports that the restoration of one of these is tough as parts simply aren't available and have to be made. *Lee Klancher*

Model 24 One-Row Picker

◀ These early one-row corn pickers were not built and sold in great numbers, and only a handful have survived.

Lee Klancher

1955 Farmall 400 and 2MH

▲ This late-model original-condition 2MH corn picker has an auto lube system, elevator fan, and metal wheel shields. It is owned by Andrew Tucker. *Lee Klancher*

1955 Farmall 400 and 2MH

▶ The operator's platform of the 2MH is a cramped space that isn't easy to slide into—a far cry from modern machinery. The 400 is equipped with a Char-lynn power steering unit. *Lee Klancher*

1964 Farmall 706 and 234

◀ The 234 corn harvester is equipped with a husking bed. The original-condition 234 corn harvester has a light kit, engine screens, and a super snoot. The machine is owned by Andrew Tucker. *Lee Klancher*

1973 Farmall 766 and 234

◀ The 766 is a diesel and the 234 is a late-model corn harvester equipped with a shelling unit. The tractor has only 2,300 hours. *Lee Klancher*

COTTON HARVESTERS

Mechanical cotton harvesters were developed as early as the 1850s, but International Harvester didn't begin to develop its own until E. A. Johnston proposed a vacuum-powered machine in 1922. That unit eventually failed, but the company purchased a design and tested that in the late 1920s and into the 1930s.

The company rolled out its first production model, the H-10-H, in 1942 and said that more than 40 years of development went into the machine. It was the first successful mechanical cotton picker to be put on the market, and one of those units affectionately dubbed, "Old Red," resides in the Smithsonian Institution.

International Harvester continued to develop that model through the 1940s, and it evolved from machines that were built on a tractor chassis to machines more like combines with their own unique chassis and power unit.

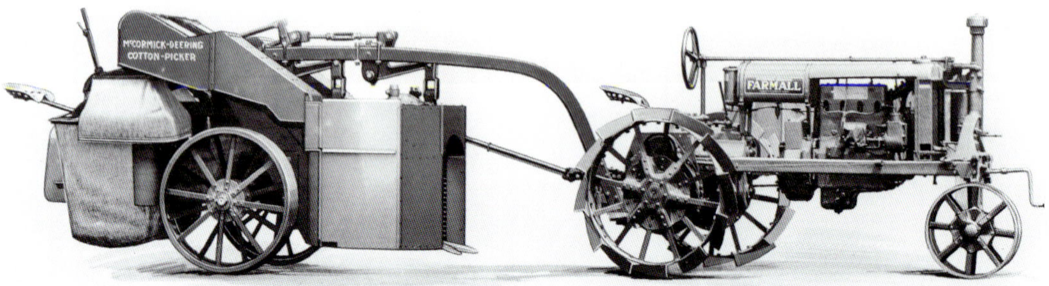

McCormick-Deering Spindle Type Cotton Picker

◄ International Harvester started developing cotton pickers quite early, as evidenced by this experimental unit photographed on August 20, 1927. The picker was designed for use in the lowlands and other sections of the Old South where the entire crop cannot be picked at one time owing to a long season and uneven ripening. The machine is equipped with two vertical picking cylinders, which carry a large number of spindles. The mechanism of the picker is operated with power taken directly from the tractor engine by a power take-off shaft. *Wisconsin Historical Society #59537*

1937 Experimental International Harvester Cotton Picker

◄ A man operating an experimental International Harvester cotton picker in a field on the Hopson Plantation near Clarksdale, Mississippi. The picker unit is raised or lowered by turning a screw near the right arm of the operator.

Wisconsin Historical Society #24584

McCormick-Deering H-10-H Cotton Picker

▲ International Harvester's cotton harvester was a major innovation introduced in 1942. The limited raw material available during and after the war limited production. An M-12-H model based on the Farmall M was introduced in 1948. Dorothy Hutchison operating a McCormick-Deering H-10-H cotton picker in a field near Firebaugh, California, in 1944.

Wisconsin Historical Society #78138

1977 International 95

▶ The 95 was a two-row picker powered by a 110-horsepower IH D310 diesel engine. *Case IH*

1978 International 782

▲ The two-row 782 was introduced in 1978. *Case IH*

1982 International 400 Cotton Stripper

▶ The 1400 was introduced in 1980. Powered by a 120-horsepower engine, the model featured the ability to harvest up to four rows into the large 637-cubic-foot basket. *Case IH*

1996 Case IH 2555

◀ The four- or five-row 2555 offers a 1,150-cubic-foot basket. The model remained in production well into the twenty-first century, and the later models were powered by a six-cylinder 260-horsepower engine. *Case IH*

2009 Case IH Cotton Express 620

◀ The 620 is a six-row cotton harvester that weighs in at 44,600 pounds and is powered by a 340-horsepower turbocharged six-cylinder engine. *Case IH*

OTHER HARVESTERS

SUGAR CANE

Case IH Sugar Cane Harvester

▲ The sugar cane harvesters are built in the Case IH plant in Piracicaba, Brazil. The machine was originally created by Austoft, a company that Case IH purchased. The factory on-board computer monitors engine performance and provides crop harvest optimization tools, including AFS. *Case IH*

Case IH 8800 Sugar Cane Harvester

▲ The 8800 is a tracked machine, and power is provided by a 353-horsepower FPT C9 engine, which can be accessed by tilting the cab to the side. The Model 8000 is the same machine with wheels. *Case IH*

Case 88-Inch Tobacco Harvester

▲ This tobacco harvester was photographed in the 1950s. *Case IH*

COFFEE

Case IH 200

◀ *Case IH*

Case IH 200

◀ *Case IH*

TU 5
HU 2

Epilogue

THE FUTURE OF THE COMBINE

By Lee Klancher

"Our twenty-first century economy may focus on agriculture, not information."
—James Howard Kunstler, author of *A History of the Future*

The rotary combine transformed the industry when it first appeared in 1977. The technology was a leap of faith, a creation that was unlikely to have emerged from a large company such as International Harvester. Innovation springs most readily from individuals and small companies free from the constraints that are inevitable as a company grows and matures.

Several key individuals believed that the rotary combine was a superior technology and worked in secret, off company budgets. The fact that they were headquartered in East Moline was an advantage in this case, as the group itself tended to set direction independently and was located far from the corporate headquarters in Chicago and also isolated a bit from the more buttoned-down confines of the engineering research center now located at Burr Ridge.

By the time the machine surfaced on the company radar in 1969, it was developed just enough to maintain momentum and continue to grow. In 1974, the discovery that New Holland was building a similar machine combined with the Brooks McCormick administration's focus on research and development caused the company to devote a large amount of resources toward making the technology a marketable machine.

The Axial-Flow combine changed the entire industry, and the machine has become a key part of Case IH's business and a valuable tool that has served tens of thousands of American farmers.

Like a large portion of the company's employees, Andreas Klauser, the president of Case IH Agricultural Equipment, grew up around farming, and he considers the Axial-Flow a cornerstone of the company. "I'm a big believer in the machine and the technology," Klauser said. He's also impressed by the longevity of the machines: "You can still see many old Axial-Flow combines that are 20 years old out and still on farms still doing a good job with the technology."

The Combine of the Future

◀ This sketch by industrial designer Gregg Montgomery was done for a 1976 display at the Kennedy Space Center in Cape Canaveral, Florida. The advanced design included many features that are now a reality, including a 24-row foldable corn header, harvesting speeds in excess of 5 miles per hour, four-wheel drive, satellite navigation, and high-floatation tires. *Gregg Montgomery Collection*

The Farm of Tomorrow

▲ As the world population continues to grow, demand for food will as well. Farmers will feed those mouths with improved techniques and technology. *Case IH*

The Axial-Flow has evolved and grown over the years, with massive improvements in efficiency, capacity, and ease of use. "It's the backbone of the Case IH agricultural business and the backbone of the rotary industry," Klauser said.

Later in life, Don Murray, who headed the engineering team during the key years of the original Axial-Flow's development, remarked that while he was surprised to see how far the technology had come, he also believed that he and many of the people around him envisioned that the technology would evolve much in the way it did.

The original rotary combine was built because of a burning belief that the technology could play a role in the industrialization of the farm, and it did just that. As the machine continues to evolve, the farmer of today will play a key role in shaping how the machine grows.

American farmers today are business people in a way that the farmers of the 1950s could hardly imagine. Farmers still made up 12 percent of the population at that time, and the average farmer worked 216 acres of land.

Today, farmers make up about 0.6 percent of the American population, and the average farmer works 434 acres of land.

Similar trends can be seen around the globe, and, while needs and conditions vary greatly by region, farmers in any

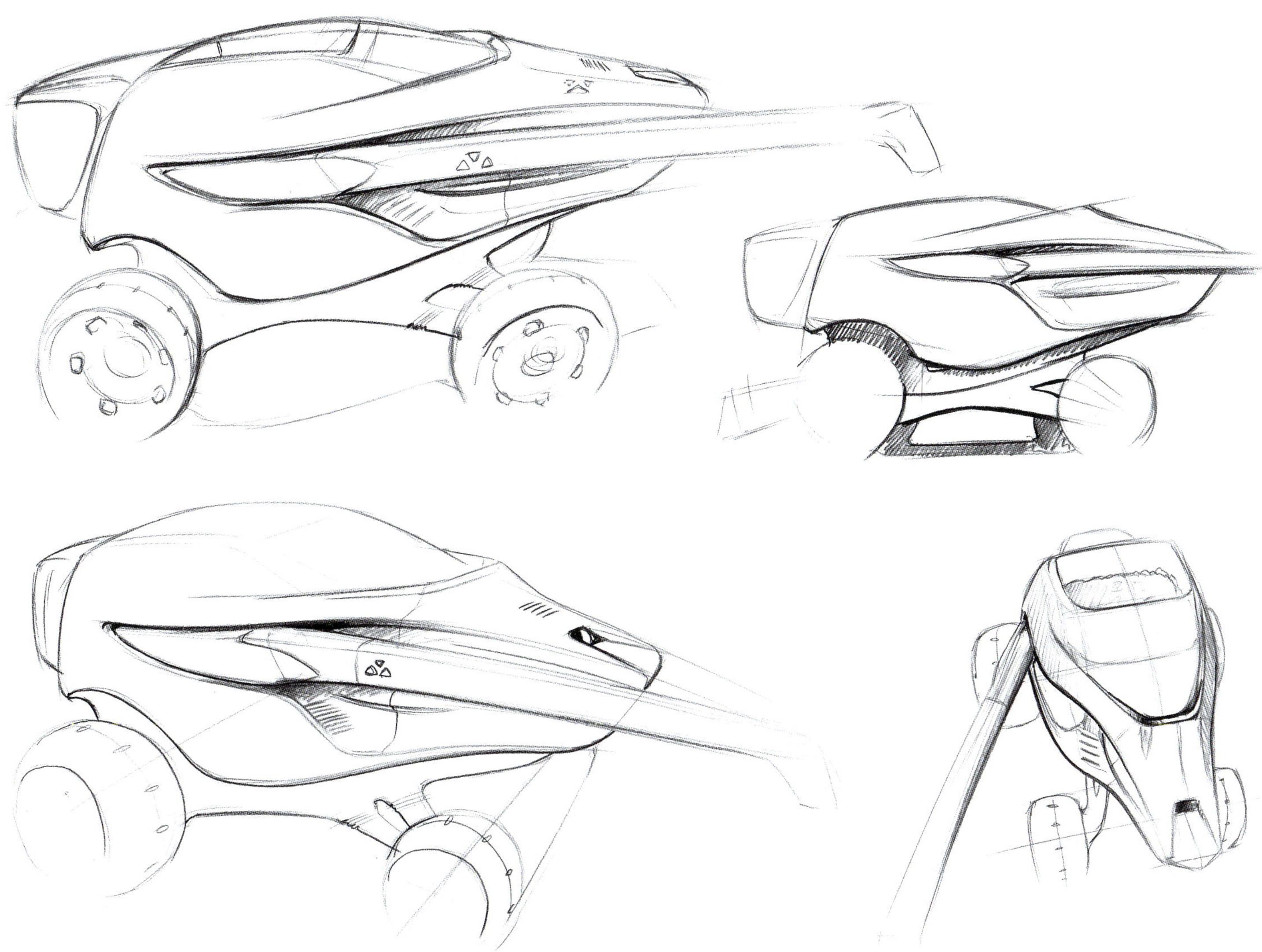

Combines Imagined

◀ The future of the combine, as envisioned by Case IH's design team. *Case IH*

corner of the world require more efficiency, productivity, and less downtime.

Klauser stated that working closely with farmers and meeting those needs are keys to developing the combine of the future. "At Case IH, we are very close to the customers and understand what they need," Klauser said. "That has allowed us to build what I believe is the most efficient combine in the industry. In the future, I believe our customers will be more successful using our technology."

He added that the company's customer support is another key focus, and that it has grown more flexible and responsive to the needs of farmers. "We will continue to have and develop excellent service and parts support all across the world, which is ready for all the regions," Klauser said. "The harvesting window is sometimes short and unpredictable, and with our machines we are carrying the user to get the most out of it.

"We don't let them down."

The Axial-Flow combine has harvested crops around the world, offering improved efficiency and a better life for farmers from Uzbekistan to Brazil to Iowa. The future is fickle, but if the past is any judge, more farmers around the planet will benefit from the technology that began with a Swedish engineer's dream.

ACKNOWLEDGMENTS

Creating this book required a team, and this team was terrific.

Co-author Gerry Salzman lived the history of the Axial-Flow combine, and his input shaped the entire project from start to finish. He loves red combines and the people who live them, and his influence on the industry was huge. So many doors opened when people understood Gerry was involved because they had positive relations with him that spanned generations. He also has an amazing nose for a good story.

Dave Gustafson's archives are amazing. The Case IH engineer appeared to save every interesting scrap of information about combines ever created.

Retired chief engineer Don Watt was also an extremely helpful resource, and his straight-shooting answers were enlightening and appreciated.

The memoir written by the late IH engineer Don Murray was vital to this book. His detailed account of the creation of the Axial-Flow was invaluable.

Thanks to all the people interviewed for the book: John Aaskov, Bill Baasch, Camiel Beert, Will Bushell, Jose Camargo, Randy Cochrane, Chet and Dan Eyer, Rick Farris, Don Fast, Red Gochanour, John Hansen, Paul Harrison, Gerald and Ed Heim, Len and Gerald Hergott, John Hinkle, Rob Holland, Charlie Hoober, Jamie and Steven Horst, Ken Johnson, Larry and Casey Jones, the late Dr. Glenn Kahle, Duane Keller, Dan Kennedy, Andreas Klauser, Kelly Kravig, Christian Lancestremere, Dave Larson, Jose Veiga Leal, Steve Lee, Edgardo Legar, Sean Lennon, Patrice Loiseleur, Jim Lucas, John Machin, Rich McMillen, Trevor Mecham, Jim Minihan, Rafael Miotto, Gregg Montgomery, Dave North, Ed Powell, Steve Puscher, Geoff Rendell, Mirco Romagnoli, Jon Ricketts, Carlos Rosa, Jay Schroeder, Eric Shuman, Tom Sleight, Derek Stimson, Rick Tolman, Steve Tyler, Jim Walker, Pedro Valente, Jose Veiga, August Von Eckardstein, Don Watt, Nate Weinkauf, Terry Wolf, and Weihong Zhang.

Thanks to Case IH licensing manager Sarah Pickett for her support, cooperation, and advocacy.

Photo researcher Gregory T. Smith did a terrific job digging up images from the McCormick–International Harvester Collection housed at the Wisconsin Historical Society. Thanks to Lorry Dunning for his research in the Hal F. Higgins Collection, and to the University of Guelph in Ontario, Canada.

Thanks to Sally Jacobs, the new archivist for the McCormick–International Harvester Collection, for her extraordinary efforts getting us the right images on time and the ability to laugh even when things don't go quite as planned.

Thanks to contributors Ken Updike, Sarah Tomac, Jean Cointe, Matthias Buschmann, Johann Dittmer, Gregg Montgomery, and Ron R. Schmitt.

Thanks to the collectors whose machines appear in the book:

Max Armstrong, Matt Frey, Brian Hale, Darius Harms, Jeff Jacobs, James M. Jecha, Marty Nigon, Josh Olson, Dan Robinson, Randy Stokosa, Dan, Debbie and Adam Tordai, Andrew Tucker, Howard Ulrich, Richard, Mary, and Stuart Wakeman, and Tim and Christine Wilhelmi. A special thanks to the Tordai family (Daniel J., Debbie, Adam, and Daniel F.) for readying a large number of cantankerous old machines in a few hardworking days.

Thanks to the good people on Red Power Forum.

Thanks to editors and proofreaders Karen O'Brien and Leah Noel for their careful work, and to designer Tom Heffron for making the book's cover and interior look great.

Thanks to John Koharski, who did what he does best for the book despite being half in the bag most of the time.

Dr. Joan Hughes is my wife and best friend, and her support, love, and guidance make all this worthwhile.

▲ This piece of art was done by artist Charles Freitag to celebrate a century of red combines. *Charles Freitag / Case IH*

Appendix

RED COMBINE MODEL LIST

Early Combines

Model No.	Years Built
Deering No. 1 pull-type	1914–1915
McCormick No. 1 pull-type	1914–1915
Deering No. 2 pull-type	1915–1921
McCormick No. 2 pull-type	1916–1921
Deering No. 3 pull-type	1923–1927
McCormick No. 4 pull-type	1924–1926
McCormick No. 5 pull-type	1925
McCormick No. 6 pull-type	1922–1924

McCormick-Deering Models

Model No.	Years Built
No. 7 pull-type	1925–1930
No. 8 pull-type	1927–1933
No. 9 pull-type	1926–1927
No. 10 pull-type	1927–1933
No. 11 pull-type	1927–1938
No. 20 pull-type	1930–1936
No. 21 pull-type	1932–1937
No. 22 stationary thresher	1934–1940
No. 22 pull-type	1935–1945
No. 28 stationary thresher	1934–1940
No. 31 spike tooth pull-type	1935–1940
No. 31 rub bar pull-type	1935–1942
No. 36 thresher	
No. 36 stationary	1942–1946
No. 41 spike tooth pull-type	1935–1940
No. 51 hillside pull-type	1935–1950
No. 60 pull-type	1937–1938
No. 61 pull-type	1939–1940
No. 42 rub bar pull-type	1940–1944
No. 52 pull-type	1943–1950
No. 62 pull-type	1941–1951
No. 102 pull-type	1938–1939
No. 103 rub bar pull-type	1935–1939
No. 103 spike tooth pull-type	1935–1939
No. 104 spike tooth pull-type	1935–1940
No. 123 SP	1942–1948
No. 125 SP	1948–1949

McCormick-IH Models

Model No.	Years Built
125SPV	1950–1951
125SPVC	1951–1952
127 SP	1952–1954
122 pull-type	1946–1949
122 C pull-type	1949–1953
64 pull-type	1950–1954
160 hillside pull-type	1951–1953
140 pull-type	1954–1962
76 pull-type	1955–1958
141 SP	1954–1957
141 hillside SP	1955–1958
101 SP	1956–1961
151 SP	1957–1961
151 hillside SP	1959–1961
181 SP	1958–1961
80 pull-type	1959–1965
91 SP	1959–1962
93 SP	1962–1968
150 pull-type	1954–1962
82 pull-type	1966–1974

International 03 Series

Model No.	Years Built
203	1963–1966
303	1962–1967
403	1962–1970
403 hillside / side-leveling	1962–1969
403 hillside / fully self-leveling	1962–1972
503	1962–1968
105 SP	1967–1970
205	1966–1972
402 pull-type	1966–1971

International 15 Series

Model No.	Years Built
315	1967–1970
815 High-Profile	1969–1973
915 High-Profile	1969–1973
615	1971–1975
715	1971–1974
914 pull-type	1971–1974
453 hillside	1973–1977
815 Low-Profile	1974–1978
915 Low-Profile	1974–1978

International Axial-Flow, 1400 Series

Model No.	Years Built
1440	1977–1985
1460	1977–1985
1480	1978–1985
1482 pull-type	1980–1986
1470 hillside	1980–1985
1420	1981–1985

Case IH Axial-Flow, 1600 Series

Model No.	Years Built
1620 (IH Engine)	1986–1988
1640 (IH Engine)	1986–1988
1660 (IH Engine)	1986–1988
1680 (IH engine)	1986–1988
1682	1987, 1990–1991
1670 hillside	1989–1991
1620 XPE (CDC Engine)	1989–1992
1640 XPE (CDC Engine)	1989–1991
1660 XPE (CDC Engine)	1989–1992
1680 XPE (CDC Engine)	1989–1992
1644	1993–1994
1666	1993–1994
1688	1993–1994

Case IH Axial-Flow, 2100 Series

Model No.	Years Built
2144	1995–1997
2166	1995–1997
2188	1998–2003

Case IH Axial-Flow, 2300 Series

Model No.	Years Built
2344	1998–2002
2366	1998–2004
2388	1998–2006
2377	2005–2006

Case IH Axial-Flow, AFX Combines

Model No.	Years Built
8010	2003–2008
7010	2007–2008

Case IH Axial-Flow, 2500 Series

Model No.	Years Built
2577	2007–2008
2588	2007–2008

Case IH Axial-Flow, 120 Series

Model No.	Years Built
7120	2009–2011
8120	2009–2011
9120	2009–2011

Axial-Flow, 88 Series

Model No.	Years Built
5088	2009–2011
6088	2009–2011
7088	2009–2011

Case IH Axial-Flow, 130 Series

Model No.	Years Built
5130	2012–2013
6130	2012–2013
7130	2012–2013

Case IH Axial-Flow, 230 Series

Model No.	Years Built
7230	2012–2014
8230	2012–2014
9230	2012–2014

Case IH Axial-Flow, 140 Series

Model No.	Years Built
5140	2014–
6140	2014–
7140	2014–

Case IH Axial-Flow, 240 Series

Model No.	Years Built
7240	2015–
8240	2015–
9240	2015–

French-Built Red Combines

Model No.	Years Built
McCormick-IH F44	1953–1959
McCormick-IH F 64	1951–1959
McCormick-IH F8-68	1960–1963
McCormick-IH F8-44	1960–1964
McCormick-IH F 141	1957–1958
McCormick-IH F8-151	1959–1962
McCormick-IH F8-83	1957–1962
McCormick-IH F8-61	1957–1959
McCormick-IH F8-63	1958–1964
McCormick-IH F8-413	1964–1967
McCormick-IH F8-513	1964–1967
McCormick-IH E8-41	1963–1965
McCormick-IH 8-41	1965–1969
McCormick-IH 8-51	1969–1972
McCormick-IH 8-61	1969–1972
McCormick-IH 8-71	1969–1972
McCormick-IH 8-91	1971–1972
IH 221	1972–1973
IH 321	1972–1981
IH 431	1972–1980
IH 531	1972–1978
IH 541	1978–1980
IH 923	1980–1984
IH 933	1980–1984
IH 943	1980–1983
IH 953	1977–1983
IH E 1420	1982–1985
IH E 1440	1982–1985
IH E 1460	1982–1985

Danish-Built Models

Model No.	Years Built
Case IH Dania 3000	
Case IH Dania D 4500	
Case IH Dania D 7200	
Case IH Dania 7500	
Case IH Dania D 8400	
Case IH Dania D 8500	
Case IH Dania D 8700	
Case IH Dania D 8900	
Case IH Dania D 9000	

UK-Built Red Combines

Model No.	Years Built
McCormick B-64	

UK-Exclusive Models

Model No.	Years Built
Case IH 2366 X-Clusive	2003–2008
Case IH 2388 X-Clusive	2003–2008

German-Built Red Combines

Model No.	Years Built
McCormick-IH D-44	1953–1955
McCormick-IH D-66	1953–1955
McCormick-IH D-61	
McCormick-IH D8-61	1957–
McCormick-IH D8-62	1957–
Case IH 514	
Case IH 515	
Case IH 521	
Case IH 527	
Case IH Arcus 2500	
Case IH CT5050	
Case IH CT5060	
Case IH CT 5070	
Case IH CT 5080	
Case IH CF50	
Case IH CF60	
Case IH CF70	
Case IH CF80	

Australian-Built Red Combines

PTO Driven	Years Built	Total Number Built
GL200	1947–1957	6715
A8-1	1959–1961	3134
A8-4	1962–1973	4920
8-6	1968–1976	1835
710	1977–1979	1080
726	1980–1982	722
Total Production of PTO Driven Headers		**18,406**

Self-Propelled	Years Built	Total Number Built
A8-2	1959–1961	218
A8-3	1961–1962	244
8-5	1966–1973	1219
711	1973–1977	2126
Total Production of Self-Propelled		**3,807**

Latin American-Built Red Combines

Model No.	Years Built
Case IH 2388	
Case IH 2799	
Case IH 2566	
Case IH 2688	

CONTRIBUTORS

Red Combines 1915–2015 was created by a team of writers, photographers, and researchers from around the world. Intensive research was done in the Case IH company archives as well as archives at UC-Davis, UW-Madison, and the University of Guelph in Ontario, Canada. More than 50 IH and Case IH current and former employees were interviewed, and hundreds of documents were referenced.

Matthias Buschmann is a former Case IH executive who has co-authored five books about German IH tractors and operates an authoritative German IH tractor web site. The industrial salesman worked for Bauer & Schaurte, the company that invented the Allen wrench. He lives near Neuss, Germany, does ongoing research on IH and Case IH equipment and literature, and is working on several additional books.

Jean Cointe is a longtime IH equipment enthusiast who has an extensive collection of IH machinery, toys, brochures and memorabilia. The Frenchman owns a 100-year-old farm equipment dealership in southeastern France.

Johann Dittmer has co-authored several books about German IH equipment. He grew up on a farm in northern Germany, and owns a large collection of tractor-related literature and images from all around the world.

Lee Klancher has been publishing his writing and photography for more than 20 years. His first book, *Farmall Tractors*, was published in 1994. Since that time, he has authored more than 20 books and hundreds of magazine articles. He has visited five continents to gather stories and images, which have appeared in *Men's Journal*, *Motorcyclist*, *Successful Farming* and many more. Klancher is the founder and publisher at Octane Press. He lives in Austin, Texas with his wife, Joan, two dogs, and a garage packed with bicycles, a battered Toyota FJ40, and an old Ducati.

Martin Rickatson is a UK-based freelance journalist specializing in farm machinery and crop production, working for magazines in the UK, Ireland, the U.S. and New Zealand, as well as providing the writing for the technical, advertising and marketing needs of farm equipment companies including Case IH. Previously he spent seven years as deputy editor of the worldwide farm machinery magazine *Profi International*. His father, grandfathers and uncle all owned IH and Case IH tractors, and he has spent many hours driving red machines from the two brands' UK and German factories, as well as collecting IH literature. His website is www.jmragriculture.com.

Gerry Salzman spent 42 years working for IH and Case IH. He retired in 2014 as Senior Director, Global Marketing, after playing key roles in product development and marketing, primarily with combines and harvesting equipment lines. He was a longtime board member with the U.S. Grains Council (USGC). In 2012, Salzman received the Lifetime Achievement award from the USGC and was only the sixth recipient in 50 years. As a graduate of the Illinois Agricultural Leadership Program he was honored with the inaugural Torch of Leadership award in 2014. He served six years in the U.S. Army Reserves, is a Alpha Gamma Rho member, and is a patent award recipient. Gerry and Judy have two children, Amanda and Adam, and four grandchildren.

Gregg Montgomery is an industrial designer who began his professional career with Caterpillar and later worked for Ford Styling and Design before joining International Harvester in 1975. Gregg managed the IH Industrial Design department before establishing his own consulting firm, Montgomery Design International, Inc., in 1983. Gregg has held design responsibility for Caterpillar's light duty product line, plus the IH and Case IH agricultural equipment product lines, and numerous products for Ford, Komatsu, New Holland, AGCO, and many more. Gregg has received various professional design awards and holds 17 patents for the wide variety of products he has designed.

Ron R. Schmitt is an areospace engineer who worked for Boeing Company and, as of 2014, was an engineering manager at Spirit AeroSystems in Wichita, Kansas. He grew up on his father's IH dealership, is an active member of the International Harvester Collectors Club, and has authored several articles for *Harvester Highlights*.

Gregory Smith is a professional research historian with more than 25 years of experience doing archival research in the Library of Congress, the National Archives, the Wisconsin Historical Society, and many more. A former US Army Special Forces paratrooper, Smith is the owner of History Piquette LLC and he resides in Wisconsin.

Sarah Tomac (formerly Galloway) is the author of *International Harvester Australia: Geelong Works*. Tomac discovered her passion for writing about IH tractors while living in Australia, and spent 10 years writing about them for enthusiast magazines. She began driving tractors at a young age on her family farm. She now resides in Chesaning, Michigan with her family.

Kenneth Updike has authored several books about International Harvester farm tractors and writes regularly about late model IH tractors for *Red Power* and *Heritage Iron* magazines. He works at a Case IH dealership in Juda, Wisconsin and lives in nearby Evansville.

INDEX

Ode to the Axial Flow Combine

By Gary Wells, IH Project Manager, late 1970s

The axial-flow is the best combine yet.
It's easier to run, easier to set.

The corn head and grain head are the best in the field.
They interchange easily and they get all the yield.

The one massive rotor will turn slow or turn fast;
And it's heavy and rugged and constructed to last.

A great big grain tank mounted on top
Will hold a full round from 'most any crop.

Electronic monitors and servo-control
Put you in command across valley, plain, and knoll.

A big diesel engine powers you through
Despite the tough going that nature handed to you.

The new rotary principle recovers all the grain,
While the super deluxe cab fends off the rain.

No straw racks to plug, no straw to pitch,
Just climb in and drive the sonofabitch.